Glutathione System and Oxidative Stress in Health and Disease

Edited by Margarete Dulce Bagatini

Published in London, United Kingdom

IntechOpen

Supporting open minds since 2005

Glutathione System and Oxidative Stress in Health and Disease
http://dx.doi.org/10.5772/intechopen.82965
Edited by Margarete Dulce Bagatini

Assistant to the Editor(s): Charles Assmann

Part of IntechOpen Book Series: Biochemistry, Volume 17
Book Series Editor: Miroslav Blumenberg

Contributors
Alejandro Cordova-Izquierdo, Eren Sarikaya, Selami Doğan, Nadège Ngombe Kabamba, Théophile Mbemba, Paulin Mutwale Kapepula, Mboloko Esimo, José Remacle, Yonca Yuzugullu Karakus, Aoula Moustapha, Figen Öngöz Dede, Margarete Bagatini, Charles Elias Assmann, Naiara Stefanello, Nathieli Bianchin Bottari, Jucimara Baldissarelli, Maria Rosa Chitolina Schetinger, Vera Maria Melchiors Morsch, Adrian Emmanuel Iglesias Reyes, Alda Roció Ortiz Muñiz, María de Lourdes Juárez Mosqueda, Jesús Alberto Guevara González, Juan Eulogio Guerra Liera, Gustavo Ruiz Lang, Rubén Huerta Crispín, Abel E. Villa Mancera, Pedro Sánchez Aparico and Raúl Sánchez Sánchez

First published in London, United Kingdom, 2020 by IntechOpen
IntechOpen is the global imprint of INTECHOPEN LIMITED, registered in England and Wales, registration number: 11086078, 5 Princes Gate Court, London, SW7 2QJ, United Kingdom
Printed in Croatia

British Library Cataloguing-in-Publication Data
A catalogue record for this book is available from the British Library

Additional hard and PDF copies can be obtained from orders@intechopen.com

Glutathione System and Oxidative Stress in Health and Disease
Edited by Margarete Dulce Bagatini
p. cm.
Print ISBN 978-1-83880-125-0
Online ISBN 978-1-83880-126-7
eBook (PDF) ISBN 978-1-83968-018-2

For EU product safety concerns:
IN TECH d.o.o., Prolaz Marije Krucifikse Kozulić 3, 51000 Rijeka, Croatia, info@intechopen.com or visit our website at intechopen.com.

IntechOpen Book Series
Biochemistry
Volume 17

Dr. Margarete Dulce Bagatini is an associate professor at the Federal University of Fronteira Sul/Brazil. She has a degree in Pharmacy and a PhD in Biological Sciences: Toxicological Biochemistry. She is a member of the UFFS Research Advisory Committee and a member of the Biovitta Research Institute. She is currently: the leader of the research group: Biological and Clinical Studies in Human Pathologies, professor of postgraduate programs in Biological Sciences: Toxicological Biochemistry at UFSM, postgraduate program in Biochemistry at UFSC and postgraduate program in Science and Food Technology at UFFS. She has experience in the area of pharmacy and clinical analysis, acting mainly on the following topics: oxidative stress, the purinergic system and human pathologies, being a reviewer of several international journals and books.

Editor of Volume 17:
Margarete Dulce Bagatini
Federal University of the Fronteira Sul, Academic Coordination,
Chapecó Campus, Chapecó, SC, Brazil

Book Series Editor: Miroslav Blumenberg
NYU Langone Medical Center, New York, USA

Scope of the Series

Biochemistry, the study of chemical transformations occurring within living organisms, impacts all of life sciences, from molecular crystallography and genetics, to ecology, medicine and population biology. Biochemistry studies macromolecules - proteins, nucleic acids, carbohydrates and lipids –their building blocks, structures, functions and interactions. Much of biochemistry is devoted to enzymes, proteins that catalyze chemical reactions, enzyme structures, mechanisms of action and their roles within cells. Biochemistry also studies small signaling molecules, coenzymes, inhibitors, vitamins and hormones, which play roles in the life process. Biochemical experimentation, besides coopting the methods of classical chemistry, e.g., chromatography, adopted new techniques, e.g., X-ray diffraction, electron microscopy, NMR, radioisotopes, and developed sophisticated microbial genetic tools, e.g., auxotroph mutants and their revertants, fermentation etc. More recently, biochemistry embraced the 'big data' omics systems.

Initial biochemical studies have been exclusively analytic: dissecting, purifying and examining individual components of a biological system; in exemplary words of Efraim Racker, (1913 - 1991) "Don't waste clean thinking on dirty enzymes." Today however, biochemistry is becoming more agglomerative and comprehensive, setting out to integrate and describe fully a particular biological system. The "big data" metabolomics can define the complement of small molecules, e.g., in a soil or biofilm sample; proteomics can distinguish all the proteins comprising e.g., serum; metagenomics can identify all the genes in a complex environment e.g., bovine rumen. This Biochemistry Series will address both the current research on biomolecules, and the emerging trends with great promise.

Contents

Preface

Oxidative stress, a state that reflects an imbalance between the production of reactive oxygen species (free radicals) and antioxidant defenses, is known to play a pivotal role in the development and progression of several chronic human diseases. To prevent damage to important cellular components (such as DNA, proteins, and lipids) biological systems present enzymatic and nonenzymatic defenses to counteract high levels of free radicals besides the antioxidants obtained from the diet. Important lines of defense against the harmful effects caused by reactive chemical species include: superoxide dismutase, catalase, enzymes of the glutathione system (such as glutathione peroxidase) and also glutathione, the most abundant thiol in animal cells.

The goal of this book is to present and discuss in more detail some topics especially related to, but not limited to, the glutathione antioxidant system in health and disease. Hopefully, this book will be a handbook for researchers and students providing basic and advanced knowledge and encouraging future research in related fields. The book consists of seven chapters, covering some important aspects of glutathione and glutathione antioxidant enzymes, but also including information about the role of oxidative stress and other important antioxidant enzymes, and the function of the immune and purinergic systems in the modulation of the glutathione system. In this sense, this book provides interdisciplinary information and within each chapter, the reader will be given an overview of the subject to be covered.

A brief outline of the book chapters is as follows. Chapter 1 (*Modulation of glutathione antioxidant enzymes in health and disease: Role of the purinergic and immune systems – Charles Elias Assmann*) discusses the interplay between the purinergic and immune systems in the modulation of the glutathione system in health and disease. Chapter 2 (*Subcellular localization of glutathione peroxidase, change in glutathione system during ageing and effects on cardiometabolic risks and associated diseases - Théophile Mbemba Fundu*) addresses the localization of glutathione peroxidase in cellular compartments, the alterations in its activity in the process of cellular aging, and the impact on cardiometabolic risk and associated diseases. Chapter 3 (*GPx activity and diseases - Eren Sarıkaya*) provides an overview of the role of glutathione peroxidase in diseases. Chapter 4 (*Neurodegenerative diseases: Potential effect of glutathione - Aoula Moustapha*) presents some insights into the role of glutathione in neural cells and discusses some consequences of glutathione disturbances in neurodegenerative pathologies, such as Alzheimer's and Parkinson's disease, in which oxidative stress plays a major role. Chapter 5 (*Periodontal health and disease in glutathione - Figen Öngöz*) presents information on glutathione peroxidase in periodontal health and disease, including aspects related to systemic disease, smoking, wound healing, and medication. Chapter 6 (*Effect of oxidative stress on sperm cells - Alejando Córdova Izquierdo*) examines some general aspects of oxidative stress in cells and its effects in sperm cells, also reviewing the effects on male and female fertility. And finally Chapter 7 (*Typical catalases: Function and structure - Yonca Yuzugullu Karakus*) focuses on the function and structure of mono-functional heme-catalases, emphasizing the information obtained in the last few years mainly in relation to the secondary activity of these enzymes.

In conclusion, based on the broad range of issues covered in this book, we hope that both academic researchers and students can learn and share the knowledge presented here. This book is the result of several collaborating parties. We gratefully acknowledge all the authors and reviewers for their precious contribution to this book, the valuable assistance of Ms. Lada Bozic, and all the support of the book publisher throughout the publication process.

Margarete Dulce Bagatini
Federal University of the Fronteira Sul,
Academic Coordination,
Chapecó Campus,
Chapecó, SC, Brazil

Section 1

Glutathione System in Health and Disease

Chapter 1

Crosstalk between the Purinergic and Immune Systems: Implications for the Glutathione Antioxidant System in Health and Disease

Charles Elias Assmann, Naiara Stefanello,
Nathieli Bianchin Bottari, Jucimara Baldissarelli,
Maria Rosa Chitolina Schetinger,
Vera Maria Melchiors Morsch and Margarete Dulce Bagatini

Abstract

Glutathione (GSH) represents the major nonprotein thiol in cells and, alongside with glutathione-dependent enzymes such as glutathione reductase (GR), glutathione peroxidase (GPx), and glutathione S-transferase (GST), exerts several biological functions including the protection against free radicals and other essential metabolic reactions within the body. Disturbances in the homeostasis of this complex glutathione antioxidant system may damage cells and have been implicated with the development and progression of several human diseases. In this context, the immune and purinergic systems are also essential, since the dysregulation in both systems may also be correlated with numerous diseases. These two networks are closely related and control inflammatory responses, especially by the crosstalk of signaling molecules, receptors, and enzymes; thus, they can exacerbate or slow down the progression of diseases. Based on this background, we aimed to provide a general scenario of the purinergic and immune systems and the connection between both and the modulation of glutathione and glutathione-dependent enzyme expression and activity in the context of health and disease.

Keywords: ectonucleotidases, receptors, cytokines, inflammation, reactive species, oxidative stress, signaling, activity, metabolism

1. Introduction

Glutathione (GSH) represents the major nonprotein thiol in cells and exerts numerous biological functions including the defense against reactive oxygen species and reactive nitrogen species (ROS and RNS, respectively). Moreover, GSH participates in countless cellular and metabolic processes in the body, and, thus, changes in its homeostasis can cause irreversible cellular damage and influence the etiology and evolution of several human diseases, such as cardiovascular, inflammatory, neurodegenerative, and metabolic diseases and cancers, among others [1, 2].

However, in addition to the direct action of GSH, a second line of defense against oxidation is carried out by glutathione-dependent enzymes that counteract the negative effects of free radicals. These enzymes are (a) glutathione reductase (GR), which regenerates reduced GSH from its oxidized form, glutathione disulfide (GSSG); (b) glutathione peroxidase (GPx), which detoxifies the cell from organic and inorganic peroxides; and (c) glutathione S-transferase (GST), which catalyzes the conjugation of GSH with diverse compounds that are produced in the presence of oxidative stress detoxification [3].

Oxidative stress and inflammation are closely linked, and cells elicit antioxidant defenses against free radicals, as well as pathogens, and other foreign substances by activating immune responses [4]. This results in a sophisticated interaction between immune system cells and several molecules released by them to defend the organism against microorganisms or damaged cells from injured tissues and maintain tissue homeostasis [5]. Nevertheless, when the inflammatory responses are exacerbated and the mechanisms of homeostatic control do not work properly, this may trigger further tissue damage which is associated with several diseases [6]. Furthermore, purinergic signaling, comprised by an intricate network of receptors, enzymes, and signaling molecules, has been shown to participate in numerous cellular functions in the context of health and disease, especially immunomodulatory functions, since the components of the purinergic system are widely expressed in immune cells of several tissues [7–9].

In this chapter, we provide a general scenario of the purinergic and immune systems and how they interplay by modulating glutathione and glutathione-antioxidant enzymes in the context of health and disease.

2. Purinergic system

The formulation of a purinergic neurotransmission hypothesis was firstly proposed by Geoffrey Burnstock back in 1972 [10]. Burnstock, in his search for answers about "what molecule could be the transmitter released during non-cholinergic/non-adrenergic inhibitory transmission in the gut," suggested that perhaps adenosine triphosphate (ATP) could fill the criteria based on the following conditions needed by a neurotransmitter: (a) the substance must be present within the presynaptic neuron; (b) it must be inactivated by ectoenzymes and/or neuronal uptake; (c) it must be released by a Ca^{2+}-dependent mechanism; and (d) specific receptors for the substance must be present on the postsynaptic cell [10–12]. Although some other researchers had already highlighted the role of purines in blood vessels and the heart, and the action of ATP in the autonomic ganglia, the ATP molecule had its role as a neurotransmitter discredited in the beginning [11].

Nowadays, however, the existence of a purinergic signaling system is well-accepted and widely studied because its constituents are found in all tissues of the body and associated with immune, nervous, cardiac, hepatic, renal, metabolic, and digestive functions, among others. Besides, the purinergic system shows all the criteria needed for ATP to be considered a neurotransmitter. In the following sections, the purinergic system components will be discussed in more detail.

2.1 Nucleotides and nucleosides

Nucleotides have three characteristic components: a nitrogenous base (containing nitrogen), a pentose (sugar), and one or more phosphates. The molecule without the phosphate group is called nucleoside. Nitrogen bases are derived from two related compounds, pyrimidine and purine. The purine bases are adenine

(A) and guanine (G), and the pyrimidine bases are cytosine (C), thymine (T), and uracil (U). Nucleotides and nucleosides have important roles described in the literature such as (a) energy currency in metabolic reactions; (b) chemical bonds in cellular responses to hormones and other extracellular stimuli; (c) components of an ordered structure of enzymatic cofactors and metabolic intermediates; and (d) components in DNA and RNA structures. However, although these are the most known properties, purine and pyrimidine nucleotides have other signaling functions described below [13].

The biological properties of pyrimidine and purine nucleotides and nucleosides are mainly linked with their binding (or not) to specific receptors. Uridine nucleotide is known by its action in the metabolism of carbohydrates as uridine diphosphate (UDP)-glucose and glycogen synthesis. Besides, cUMP is related alongside with cCMP as intracellular second messengers. Cytidine is known to form cCMP, which has been associated with the control of cell growth and blood cell function; however, the intracellular signal transduction pathways are not well-defined [14]. Some authors proposed that its mechanism is related to the use of the cGMP signal transduction pathway [15]. Furthermore, cytidine (as cytidine triphosphate (CTP)) and uridine (which is converted to uridine triphosphate UTP and then to CTP) contribute to brain phosphatidylcholine and phosphatidylethanolamine synthesis [16]. Thymidine and its associated nucleotides have a role as modulators of active anticancer drugs, especially antimetabolites [17].

Guanosine and adenine nucleotides are the most commonly known. For example, GTP and cGMP are associated with intracellular signaling in physiological events of hormonal regulation; GTP and ATP (as well as CTP) are involved with the regulation of allosteric enzymes [18]. It is known that the suppression of GTP concentrations could be related to the invasion of melanoma cells and cells from other cancer types [19, 20]. Moreover, guanosine nucleotides have a role in immune response, cardioprotection, and memory formation, among others [18, 21, 22]. Adenine nucleotides, besides being part of the energy metabolism (mainly ATP), are also neurotransmitters or signaling molecules that act in the control of cellular responses, for example: (a) While ADP stimulates platelet aggregation, adenosine (Ado) inhibits this process; (b) While ATP is an excitatory neurotransmitter in the central nervous system (CNS), Ado acts in neuroprotection; (c) ATP is a pro-inflammatory molecule known as a damage-associated molecular pattern (DAMP); (d) On the other hand, Ado is an anti-inflammatory molecule. Besides, Ado nucleotides are found in all tissues of the body presenting many cellular modulatory effects [23–25].

2.2 Purinergic receptors

ATP, ADP, UTP, and UDP bind to P2 receptors, while Ado binds to P1 receptors (**Figure 1**). P2 receptors are subdivided into P2X and P2Y families [26]. P2Y receptors are metabotropic G protein-coupled receptors (GPCRs), and its eight subtypes can be divided into two groups, depending on the type of G protein-coupled receptor: P2Y1, P2Y2, P2Y4, and P2Y6 are Gq protein-coupled receptors and activate the protein phospholipase Cβ, while P2Y12, P2Y13, and P2Y14 are Gi protein-coupled receptors which act by inhibiting adenylate cyclase. P2Y11 is a peculiar receptor because it is coupled to Gq and Gs and, thus, causes an increase in the intracellular levels of 3′,5′-cyclic adenosine monophosphate (cAMP) and Ca^{2+} [27, 28].

GPCRs are the largest and most diverse group of membrane receptors. The activation of a single G protein can affect the production of second messenger molecules such as cyclic AMP, diacylglycerol (DAG), and inositol 1,4,5-trisphosphate (IP3). Furthermore, there are mainly three subtypes of G proteins, Gs, Gi,

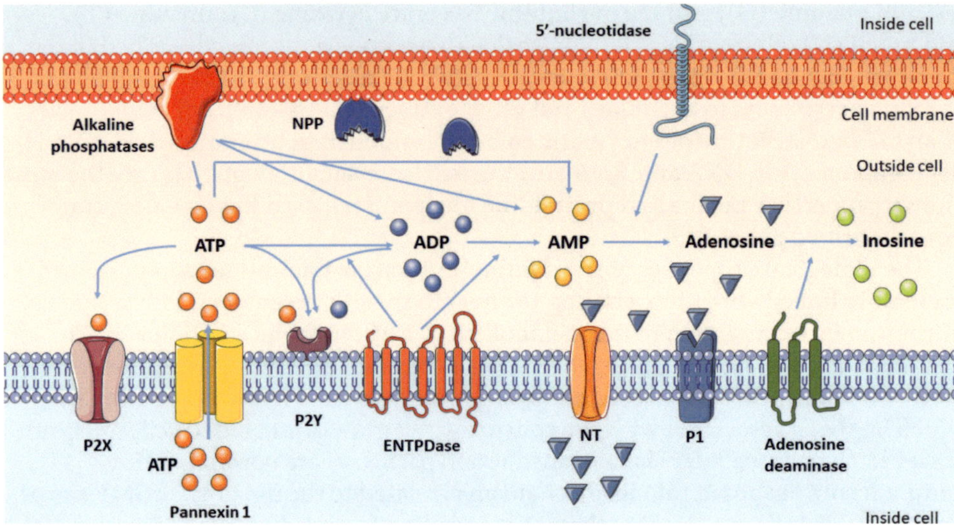

Figure 1.
Purinergic system components. P1 and P2 receptors, enzymes: ENTPDase, alkaline phosphate, NPP, 5'-nucleotidase, and adenosine deaminase. Nucleotides and adenosine. Nucleoside transporter (NT) and pannexin 1 (channel is an integral component of the P2X/P2Y purinergic signaling). (Authors' artwork).

and Gq. The Gs (stimulatory) protein activates adenylate cyclase, which catalyzes the formation of cAMP from ATP, being involved in the signaling of many receptors such as glucagon, epinephrine, and calcitonin, among others. The Gq protein is involved in the activation of the phospholipase C (PLC) enzyme that participates in the formation of second messengers. Once activated, it degrades phosphatidylinositol 4,5-bisphosphate (PIP2) present in the membrane into IP3, and 1,2-DAG. Gp protein has important functions in the brain, such as neuronal transmission, synaptic plasticity, and neuronal survival. Taking this information into account, studies have shown that Gq protein plays an important role in the processes of neurodegeneration in Alzheimer's disease. The Gi (inhibitory) protein inhibits the activity of adenylate cyclase enzyme. The Gi isoform, related to the decrease in cellular response, is responsible for mediating the inhibitory effects of receptors [29, 30]. The biological effects mediated by PY2 purinergic receptors are associated with these types of G proteins.

P2X receptors are ionotropic receptors linked to channels in the plasma membrane; they have ATP as an agonist. There are seven different P2X receptor subtypes (P2X1–7) that form trimeric receptors [27]. The extracellular domain of these receptors contains binding sites for ATP, competitive antagonists, and modulatory metal ions. The transmembrane domains form a nonselective cation channel. The opening properties of ion channels differ greatly according to the receptor subtype: while the homomeric P2X2, P2X4, and P2X7 receptors exhibit slow desensitization, the P2X1 and P2X3 receptors exhibit rapid desensitization [31].

P1 receptors, which have Ado as an agonist, are divided into four subtypes, A1, A2A, A2B, and A3, and all are G protein-coupled receptors. The A1 receptor mediates signaling responses which may be caused by its coupling to different proteins within the Gi/o family. The known pathway of action of this receptor is through the inhibition of adenylate cyclase, which causes a decrease in cAMP. Moreover, the A1 receptor mechanism of action is through the activation of PLC, which leads to an increase in cytosolic calcium. In turn, A2A receptors are Gs protein-coupled receptors and activate adenylate cyclase. A2B receptor is coupled to different signaling pathways, including guanylate cyclase activation, through PLC-mediated Gq coupling, and an increase in Ca^{2+} concentrations is dependent on IP3. A3 receptors are Gi protein-coupled and also Gq protein-coupled. A2B and A3 receptors are known as methylxanthine-insensitive receptors on the contrary of A1 and A2A receptors [26, 32, 33].

P1 receptors are widely distributed in metabolically active tissues, such as the pancreas, liver, and adipose tissues; they are also present in immune cells, indicating an important role in the regulation of immune system responses. The A1 receptor is an oxidative stress sensor, and it has shown to have pro- and anti-inflammatory effects as well as it is associated with the reduction of ischemic events [34, 35]. Moreover, A1 receptor activation has been shown during the initial phase of leukocyte recruitment, and the A2A receptor is expressed at the resolution phase [36]. On the other hand, the activation of the A2A receptor exacerbates neuronal damage as well as recruitment and activation of microglia in the CNS [35], although some anti-inflammatory effects by its stimulation in immune cells have been suggested [37]. A2B receptor has pro- and anti-inflammatory effects. Studies have demonstrated that A2B receptors stimulate pro-inflammatory cytokines like IL-4, IL-6, IL-8, IL-13, IL-19, and others; it has also been shown to activate human mast cells being involved in allergic and inflammatory disorders [33, 34]. A3 receptor is widely distributed in the immune tissue, and its functions are related to the release of allergic mediators including histamine by mast cells, suggesting a role in inflammation. In the brain, A3 and A1 receptors seem to be associated, as hippocampal A3 receptors have been shown to desensitize A1 receptors [38]. Moreover, A3 receptors were found to act in cardiovascular protection [33].

Both P1 and P2 receptors are located in several tissues, mainly the pancreas, vascular system, CNS, liver, kidney, and immune cells [39–42]. In the CNS, the expression of these receptors has been shown in many structures and proposed to be associated with the development of several pathologies, such as Alzheimer's disease [41, 43]. P2Y receptors, especially, are related to platelet function and thrombus promotion [42]. Platelets are known to express P2X1R, P2Y1R, and P2Y12R, where ADP signaling predominates through the activation of P2Y1R and P2Y12R, which is critical for initiating platelet aggregation [44, 45]. The P2X7 and P2X4 receptors are present in the kidney, and their expression can be increased in pathologies involving inflammatory processes of this tissue [46]. In the liver, the purinergic system is involved in the physiological regulation and also plays a role in the pathological processes of liver disease. Purinoceptors are also involved in bile secretion and glycogen and lipid metabolism. Moreover, the activation of P2Y1 receptors in human and rodent hepatocytes stimulates the glycogen phosphorylase enzyme [40].

It is important to highlight that when high concentrations of ATP bind to the P2X7 receptor, it can form pores in the membrane promoting inflammasome activation in macrophages and endothelial cells and, subsequently, promoting the release of cytokines, such as interleukin-1β (IL-1β), through a caspase-1-dependent process [47]. IL-1β is associated with autoinflammatory diseases as well as other inflammatory conditions such as hypoxia and hemorrhage. Furthermore, IL-1β causes a marked increase in the expansion of naïve and memory $CD4^{+}$ T cells in response to antigens and particularly when used with lipopolysaccharide (LPS) as a costimulant [48, 49].

Purinergic receptors, completely unknown 50 years ago, are nowadays widely studied as they participate in the modulation of many physiological processes and since their up- or downregulation is associated with many diseases. Besides, several nucleotides or nucleosides bind to these receptors and trigger their correspondent effects, inhibiting or stimulating downstream pathways. Furthermore, there are also some enzymes whose function is to control the levels of these molecules, which will be presented below.

2.3 Ectonucleotidases

The ectonucleotidases are divided mainly into four gene families, which include pyrophosphate/phosphodiesterases (ENPPs), alkaline phosphatases, ectonucleoside triphosphate diphosphohydrolases (ENTPDases), and 5′-nucleotidases. ENPPs act on triphosphate nucleotide (ATP and UTP) hydrolysis into monophosphate nucleotides (AMP and UMP) and pyrophosphates. Seven enzymes are found in the ENPP family. Two isoforms are capable of hydrolyzing ATP, especially isoforms ENPP1 and ENPP3. Moreover, ENPP4 is involved in ADP hydrolysis in platelets, and the other isoenzymes hydrolyze phosphodiester bonds into phospholipids. Furthermore, ENPP1 is related to bone mineralization and tissue calcification and has been described for acting on insulin resistance in diabetic patients [50, 51]. NPP2 is expressed mainly in the brain, lung, kidney, endothelial cells, and also biological fluids, being associated with intracellular modulation through its binding to activated integrins on the target cells [52].

Alkaline phosphatases have a wide substrate specificity for different phosphomonoesters and other compounds containing phosphate, including adenine nucleotides, inorganic polyphosphates, and pyrophosphates. Three isoenzymes are tissue-specific and have 90–98% homology, which are the alkaline phosphatases of the intestine, the placenta, and those of germ cells. The last isoenzyme, tissue-nonspecific alkaline phosphatase (TNAP), is approximately 50% identical to the others, and it is expressed mainly in the bones, liver, and kidney. TNAP is mostly known for its function in bone tissue mineralization [51, 53].

ENTPDases hydrolyze di- and triphosphate nucleotides into mononucleotides and inorganic phosphates. For their activity, they require Ca^{2+} and Mg^{2+} as cofactors. Eight different genes encode members of the ENTPDase family which differ in substrate specificity, cell location, and tissue distribution [53]. Four of them (NTPDases 1, 2, 3, and 8) are present on the extracellular surface of the membranes. NTPDases 5 and 6 exhibit cytoplasmic location, while NTPDases 4 and 7 are entirely located intracellularly, facing the lumen of cytoplasmic organelles [51]. Members of the membrane-bound NTPDase family show molecular masses of approximately 70–80 kDa, and they are proteins with glycosylated residues. They show sequence homology in special regions called "apyrase-conserved regions," which are important for the catalytic activity. These enzymes may exist either in monomeric or in oligomeric states constituted by two transmembrane domains close to the amino and carboxyterminal groups [51, 54].

Concerning their catalytic activity, different isoenzymes have different substrates affinities. NTPDases 1 and 2 have a preference for hydrolyzing adenine nucleotides in the detriment of uracil nucleotides. All membrane-bound NTPDases hydrolyze ATP more quickly than ADP. NTPDase 1 is the enzyme that has more affinity for ATP; however, it hydrolyzes the ADP product to AMP in the same proportion [55]. NTPDase 2 has a great preference for ATP hydrolysis. NTPDases 3 and 8 hydrolyze ATP and UTP in a similar proportion [56]. Intracellular enzymes differ in substrate preference. NTPDases 4, 5, and 6 preferentially hydrolyze NTP and NDP, but to a lesser extent ATP and ADP. NTPDases 5 and 6 prefer to hydrolyze ATP, but not ADP, while NTPDase 7 preferentially hydrolyzes UTP, CTP, and GTP, but has a very low affinity for ATP [53, 57].

The family of NTPDases also differs in their tissue location. NTPDase 1 is mainly located in immune cells, for example, lymphocytes, monocytes, and blood vessel endothelial cells, and in the CNS [58–60]. NTPDase 2 is also expressed in blood vessels and neuronal progenitor cells [53, 58]. Both NTPDases 1 and 2 are expressed in pancreas acinar cells. NTPDase 3 is mainly found in subsets of neurons, epithelial cells of the kidney, the upper respiratory system, and the digestive and reproductive systems [28, 61]. NTPDase 8 has a more restricted expression, being found in the liver, kidney, and intestine [62, 63]. Regarding the location of intracellular isoenzymes, they have a wider expression, due to their control of nucleotides inside the cell [54, 64]. For instance, in the CNS, different isoenzymes are expressed by neurons, astrocytes, and microglia [65]. Besides, this variation in isoenzymes may change according to distinct brain regions [60].

NTPDases can be coexpressed with another enzyme that continues with the nucleotide hydrolysis cascade, such as ecto-5′-nucleotidase (eN—CD73, E.C.3.1.3.5). eN is an enzyme anchored to the plasmatic membrane by glycophosphatidylinositol (GPI) with its catalytic site facing the extracellular medium, but it can also be found in the soluble form [53]. Mammalian eN consists of two glycoprotein subunits linked by non-covalent bonds. Zinc and other divalent metal ions bind to the end of the N-terminal domain. This ectoenzyme belongs to a large superfamily of metallophosphoesterases that act on different substrates, such as several nucleotides, serine/threonine phosphoproteins, and also sphingomyelins [66, 67].

eN is expressed in many tissues, being more abundant in the colon, kidney, and brain and less abundant in the liver, lung, and heart [68]. In the vascular system, eN is highly expressed in the endothelia and platelets. However, in immune cells, it is only present in some subpopulations of cells [67]. Besides, in the CNS, this enzyme can be found in different structures, including the cerebral cortex, hypothalamus, cerebellum, hippocampus, and olfactory bulb, among others [60].

Also, eN hydrolyzes ribo- and 5′-monophosphate deoxyribonucleotides to their respective nucleosides. Among these nucleotides, the prominent function of eN is the hydrolysis of AMP to Ado. According to Dunwiddie and Masino [69], Ado is considered a third "purinergic messenger." This nucleoside regulates many physiological processes, particularly in tissues involved with excitatory stimuli, such as the heart and the brain, by reducing their excitatory activity [69]. Ado effects are related to fluid transport, induced tolerance to ischemia and reperfusion in the cardiovascular system, immunity, and inflammation, among others [53].

Ado levels can be regulated by adenosine deaminase (ADA) enzyme activity, and in humans, two isoforms are expressed: ADA1 and ADA2. ADA1 is more relevant in the purinergic cascade because it catalyzes the irreversible Ado and 2′-deoxyadenosine deamination into inosine and 2′-deoxy-inosine, respectively. ADA1 is widely expressed in the intestine, thymus, spleen, and other lymphoid and nonlymphoid tissues; it is also involved in neurotransmission [70]. Moreover, liver, monocytes/macrophages, and serum also contain another isoenzyme, ADA2, which can be active at sites of inflammation during hypoxia and in areas of tumor growth [67]. Studies have shown that the ADA2 structure is precisely designed to act in the extracellular environment. ADA2 fits into the new family of adenosine deaminase-related growth factors (ADGFs), which play a role in tissue growth. Besides, Kaljas et al. [72], when analyzing $CD4^+$ T-cell subsets, showed that ADA2 particularly binds to regulatory T cells expressing CD39 and lacking the receptor for ADA1 [71, 72].

Understanding the regulatory mechanisms of purinergic signaling continues to be of great importance to several diseases since the overexpression or suppression of nucleotidase activities, receptor expression, and nucleotide/nucleoside levels are known to be involved in a variety of pathologies, including cancers and inflammatory, neurodegenerative, and cardiovascular diseases.

3. Involvement of purinergic signaling in immune responses

Immune responses are the result of a complex interaction between immune cells and several soluble factors, aimed to protect the host from the invasion by microorganisms or to eliminate apoptotic cells at sites of tissue injury, thus maintaining tissue homeostasis [5]. However, an intense inflammatory response, not properly balanced by endogenous mechanisms of homeostatic control, can lead to cell and tissue damage with the production of free radicals [6]. To avoid excessive oxidative stress, cells use different mechanisms to activate the immune system including antioxidant defenses and purinergic signaling [4]. It is worth mentioning that, since its discovery, purinergic signaling has been shown to mediate a wide range of functions in health and disease, especially immunomodulation and inflammation [9].

Immune cells recognize ATP, released from dying cells and damaged tissues, as a danger signal that elicits a variety of inflammatory responses. There is evidence that, following tissue injury, purinergic signaling response may be divided into three temporal phases [4]. First, an acute phase, when ATP is rapidly released into the extracellular space from damaged or stressed cells, accumulates to high levels and has chemotactic and excitatory effects on immune cells. Second, there is a decrease in the extracellular ratio of ATP/Ado responsible to limit the extent and duration of inflammation. Third, there is a chronic phase associated with a low extracellular ratio of ATP/Ado to promote tissue remodeling [4]. In the next sections, the functional role of purines in immune cell responses and the contribution of purinergic signaling to the mechanisms of inflammation will be highlighted.

3.1 How does ATP release promote inflammasome activation?

Necrotic and apoptotic cells release ATP, which works as a find-me signal to attract macrophages to phagocytose and remove dead or dying cells, a process that involves the activation of the NLRP3 inflammasome [73]. The NLRP3 inflammasome is a protein complex involved in IL-1β and IL-18 processing that senses a variety of signals including infection, tissue damage, and metabolic dysregulation [74]. The activation of the NLRP3 inflammasome results in the assembly of scaffold components: the cytoplasmic receptor NLRP3, the adaptor protein ASC, and the effector protein caspase-1. This association leads to the activation of caspase-1, allowing the processing of pro-IL-1β and pro-IL-18 to their mature and secreted forms which are biologically active. IL-1β production is a tightly controlled process playing a pivotal role in inflammation and the recruitment of neutrophils [75].

In pathological conditions, high levels of ATP (5 mM) are passively released from necrotic cells and act as a pro-inflammatory danger signal, activating the NLRP3 inflammasome through binding to the ionotropic P2X7 receptors [76]. Thus, the extracellular ATP (eATP) leads to K^+ efflux, membrane pore formation, and ROS-driven activation [77].

3.2 Purinergic receptors play a crucial role as stimuli for chemotaxis of inflammatory cells

Activation of purinergic receptors in immune cells can elicit either positive or negative feedback mechanisms and thus can tightly regulate immune responses [78]. All P1 and P2 receptor subtypes are expressed by immune cells, in a cell type- and differentiation-dependent manner (**Table 1**).

After an infection, leukocytes are programmed to exit the circulation and move toward epicenters of infection/inflammation, guided by chemical gradients of

Purinergic receptor	Ligand	Immune cell expression	Function
A1	Ado	Neutrophils and immature DCs	Chemotaxis
A2A	Ado	Most immune cells	Anti-inflammatory responses
A2B	Ado	Macrophages, DCs, and mast cells	Promotes IL-6 and VEGF release by macrophages and DCs, and drives mast cell degranulation
A3	Ado	Neutrophils and mast cells	Reduces neutrophil chemotaxis and stimulates mast cell degranulation
P2X7	ATP	$CD4^+$ T cells, $CD8^+$ T cells, Treg cells, iNKT cells, macrophages, and DCs	Activation of effector T cells, Treg cells, iNKT cells, monocytes, macrophages, and DCs
P2Y2	ATP/UTP	Phagocytes, DCs, monocytes, and lymphocytes	Chemotaxis and activation
P2Y6	UDP/UTP	Monocytes, macrophages, neutrophils, and lymphocytes	Activation

Adapted from Cekic et al. [4].

Table 1.
Principal purinergic receptors of immune cells: expression and functions.

different stimuli. Neutrophils are the most abundant leukocytes in the circulation, representing the first line of defense in the innate response. Neutrophils are characterized by a large phenotypic heterogeneity and functional versatility, placing these cells as important modulators of inflammatory responses [5]. Under adverse conditions, neutrophils release ATP via connexin or pannexin 1 hemichannels, and ATP undergoes rapid conversion to Ado via the CD39/CD73 axis expressed on the neutrophils surface [79].

Regarding P1 type of receptors, A1 and A3 receptors facilitate neutrophil chemotaxis, in part, by upregulating the neutrophil adhesion to tissue injury [80]. In particular, the stimulation of A1 receptors induces ROS production from activated neutrophils favoring bactericidal functions, whereas the activation of A2A receptors downregulates ROS generation [81]. Regarding P2 receptors, P2X1 receptors also mediate neutrophil chemotaxis negatively regulating systemic neutrophil activation, thereby limiting the oxidative response, coagulation, and organ damage [82]. However, P2Y11 receptors could retain the immune functions of neutrophils and reduce the injurious effects of increased neutrophil longevity during inflammation [83].

Phagocytes, such as macrophages, are innate immune cells that play an integral role in the defense of the host due to their ability to recognize, engulf, and kill pathogens while sending out danger signals via cytokines to recruit and activate inflammatory cells [84]. The P2X7 receptor has been suggested to play an important role in ATP-induced inflammation because it is mainly expressed on inflammatory cells. Furthermore, the role of P2X7 in the protection against neutrophil apoptosis has been reported as well as its association with the generation of ROS [85].

Some studies demonstrated the involvement of the P2X7 receptor in several responses of macrophages to danger, in particular the proinflammatory response mediated by IL-1β secretion, bacterial killing, and the associated macrophage death. ATP was shown to promote the maturation and release of IL-1β from macrophages, via P2X7 receptors [84]. Despite the dominant role of P2X7 in macrophages, evidence has supported the role of additional receptors. For example, the P2Y2 signaling on macrophages contributes to the clearance of apoptotic cells and also mediates the potentiation of prostaglandin E2 release involved in the induction of nitric oxide synthase (NOS) [86]. On the other hand, A2B receptor activation by Ado was reported to inhibit tumor necrosis factor (TNF)-α expression from macrophages, whereas it potentiates NOS and interferon (IFN)-γ expression contributing to the inflammatory profile.

Like macrophages, dendritic cells (DCs) are professional antigen-presenting cells (APCs), whose main role is to activate adaptive immunity (second defense line), thereby maintaining immune homeostasis and tolerance [5]. DCs express almost all known P2 receptors; besides, extracellular nucleotides exert multiple effects on these cells ranging from chemotaxis to control of cytokine release and induction of cell death [87]. Mature DCs mainly express A2A and A2B receptors, which have pro-inflammatory effects on these cells [88–90].

Together, macrophages and DCs are APCs responsible for the cell-mediated immune response and interaction with T lymphocytes [5]. T cells recognize antigens through their T-cell receptors (TCR), located at the immune synapse, and physically interact with peptides that are presented on major histocompatibility complex (MHC) molecules by APCs. This immune interaction causes the activation of T-cell receptors on lymphocytes, therefore eliciting T-cell differentiation, cytokine production, and cytotoxic activity. Once activated, T cells orchestrate effector immune cell function by recruiting macrophages, neutrophils, eosinophils, and basophils to sites of infection and inflammation and by increasing the microbicidal activity and cytokine and chemokine production of these cells [8].

T cells express many members of the P2X, P2Y, and P1 receptor families, as well as the ENTPD1 ectonucleotidase. Purinergic signal amplification in T cells occurs mainly through P2X1, P2X4, and P2X7 receptors [8]. A2A receptors are the most important receptors in regulating lymphocyte activation, where the overall effect is suppressive [91]. A2A receptors inhibit both IL-4 and IFN-γ production by both naïve $CD4^+$ T cells and Th1 and Th2 cells.

However, ATP is known to boost the activation of T cells by amplifying the TCR-induced activation and by increasing IL-2 production by P2X1 and P2X4 receptors [92]. Thus, T cells promote strong positive purinergic feedback mechanisms, which are further amplified in the confined space of the synaptic cleft. Confinement of ATP in the immune synapse results in a powerful autocrine feedback mechanism that facilitates the signal amplification required for antigen recognition (**Figure 2**) [8].

3.3 Role of Ado as a regulator of immune responses

In general, Ado has opposite effects on inflammation compared to ATP, essentially acting as an anti-inflammatory molecule [23]. Ado, for instance, inhibits adhesion to endothelial cells, reduces superoxide anion production by neutrophils, and lowers the release of pro-inflammatory cytokines (**Figure 2**) [93]. Besides, Ado facilitates the release of IL-10, an anti-inflammatory cytokine, from monocytes [94]. Ado also induces the production of vascular endothelial growth factor (VEGF), a potent inducer of angiogenesis and vascular permeability through its binding to the A2 receptors [93].

The role of Ado in regulating macrophage activation indicates that this molecule, by activating A2A, A2B, and A3 receptors, inhibits the production of several pro-inflammatory mediators such as TNF-α, IL-6, IL-12, nitric oxide (NO), and macrophage inflammatory protein (MIP)-1α by macrophages [95]. In parallel, extracellular Ado promotes the release of the anti-inflammatory cytokine IL-10 by monocytes and macrophages via A2A and A2B receptors exerting an anti-inflammatory effect. Moreover, Ado inhibits Th1 and Th2 differentiation by decreasing T-cell proliferation and IL-2 production [93, 95].

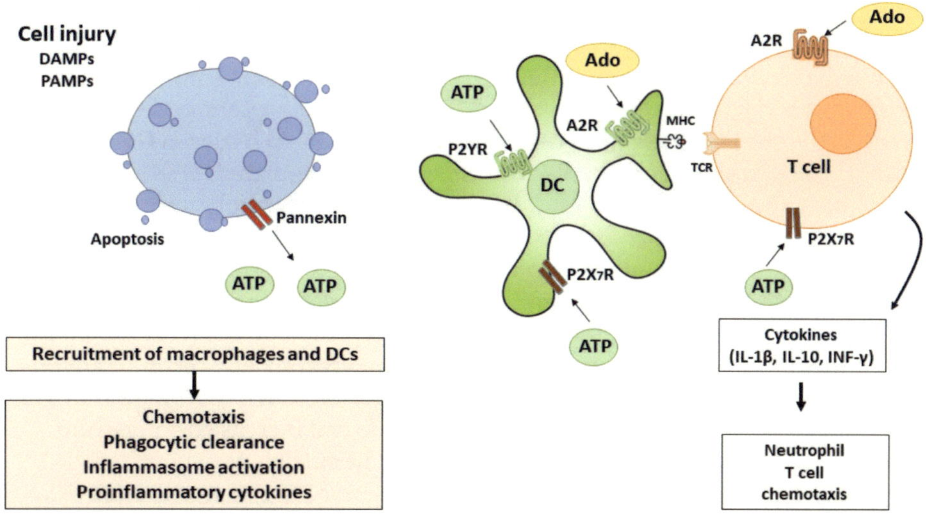

Figure 2.
Purinergic signaling during inflammation. After cell injury by DAMPs or pathogen-associated molecular patterns (PAMPs), apoptotic cells release ATP and other nucleotides by pannexin channels. ATP acts as "signal damage" for the recruitment of macrophages and DCs. Extracellular ATP ligates and activates the P2 purinergic receptors (P2Rs) and is then degraded by soluble and plasma membrane ectonucleotidases to generate ado, which acts at ado receptors (AR). DCs, as antigen-presenting cells, present antigens to the MHC of TCR, which promote the release of cytokines such as interleukin-1β (IL-1β), interleukin-10 (IL-10), and interferon-gamma (INF-γ), which then recruit neutrophil and T lymphocytes to injured tissues. Ado binds to A2 receptors and suppresses inflammation. (Authors' artwork).

As described above, inflammation appears to be oppositely regulated by extracellular ATP and Ado. At the initiation of inflammation, there are high levels of ATP, produced by damaged and stressed cells. High ATP levels promote the rapid migration of dendritic cells and macrophages through the activity of pannexin 1 channels and P1 and P2 receptors that trigger NLRP3 inflammasome activation.

Phagocytes and lymphocytes are recruited by chemoattractants and danger signals released from inflamed sites upregulating phagocytosis and other phagocyte-killing mechanisms, resulting in the clearance of the dying cells. At the end of the inflammatory process, increasing levels of Ado are induced by the breakdown of ATP by the ATP-dephosphorylating enzymes and the production of Ado by cells at the inflammatory site. This results in increased Ado levels and, consequently, inhibits the inflammatory processes.

Therefore, purinergic signaling represents the result of the activity of a complex and heterogeneous "molecular machinery" comprising nucleotide/nucleoside molecules, plasma membrane P1 and P2 receptors, and nucleotide-degrading enzymes, such as CD39 and CD73, cooperating in the inflammatory microenvironment and protecting tissues, particularly from immune-mediated excessive tissue damage.

4. The interplay of the purinergic and immune systems in the modulation of glutathione antioxidant enzymes

Besides knowing the importance of GPx, GST, and GR in several diseases, studies have drawn specific attention to the relationship between oxidative stress and purinergic signaling. Therefore, the connection between the activity and expression of glutathione antioxidant enzymes with the purinergic system is highlighted. Although many studies showed the relationship between the purinergic system and oxidative stress, they did not directly assess the activity of glutathione-dependent enzymes. In this sense, the outcomes in glutathione and purinergic system modulation of some studies addressing different human diseases were summarized in **Table 2**.

One of the studies that addressed the direct relationship between GPx and the purinergic system showed that Ado administration upregulated GPx-1 expression and activity in endothelial cells [96]. It has been shown that polymorphisms in GPx and GST enzyme genes are related to increased risk of developing coronary heart disease and stroke and is associated with elevated inflammatory markers and increased risk of coronary heart disease in smokers, respectively [97]. Moreover, in another study, the administration of an Ado receptor agonist increased GPx and GR activities in the heart of rats, and the treatment with an Ado receptor antagonist blocked this augmentation, confirming the effect of Ado on glutathione antioxidant enzymes [98]. As the generation and release of Ado can increase during acute myocardial ischemia, its high concentrations may be sufficient to induce GPx-1 expression, which results in enhanced cellular tolerance to reactive species (RS) and contributes to the cardioprotective role of Ado [96]. In agreement with this, a study performed with acute myocardial infarction patients revealed an increase in NTPDase activity, with enhanced hydrolysis of adenine nucleotides, which promotes an increased Ado generation [99].

Concerning metabolic diseases, GSH levels in erythrocytes and blood plasma were observed to be lower in patients with diabetes or metabolic syndrome (MetS), and, consequently, the reduced GSH levels potentiate the effects caused by RS [1]. Recent studies have also demonstrated the depletion of the GPx and GST enzymes in the liver [100] and heart [101] from rats with MetS. Additionally, Martins et al. [102] revealed that subjects with MetS present an increase in NTPDase, 5′-nucleotidase, and NPP activities while decreased ADA activity in platelets. Moreover, an increase in ATP and ADP hydrolysis and a decrease in Ado deamination in lymphocytes of MetS patients were observed [103]. Further, Madec et al. [104] showed an increase in the P2X7 purinergic receptor in human adipocytes, which modulates the release of inflammatory cytokines and might contribute to the subclinical inflammatory status found and conferring increased cardiovascular risk.

In the same way, Cardoso et al. [105] showed an increase in the nucleotide hydrolysis, indicating an augment in NTPDase, 5′-nucleotidase, and ADA activities in platelets from hypertensive rats, suggesting that hypertension increases Ado generation, which acts through A2A receptors. Recently, an increase in NTPDase 1, NTPDase 3, and CD73 expression and activity in the cortex and in A2A expression in the hippocampus and cortex in hypertensive rats was also demonstrated [106]. Additionally, it has been shown that animals with hypertension induced by 1,3-dipropyl-8-sulfophenylxanthine (DPSPX), an antagonist of Ado receptors, present a redox dysfunction in the initial phase of hypertension, which may be explained by the blockade of Ado's protective effects and increased generation of RS. With the interruption of DPSXP administration, Ado seems to be involved in the adaptive response to enhance the activity of vascular antioxidant enzymes, such as GPx to counteract the increase in RS generation [107].

The purinergic network emerges as a central player in pathophysiological conditions particularly linked to immune system regulation including diabetes. It has been demonstrated that Ado affects insulin secretion, glucose homeostasis, and lipid metabolism through the activation of four Ado receptors [108]. In this context, studies demonstrated that activities of enzymes that hydrolyze adenine nucleotides and nucleosides were changed in diabetic rats [109, 110]. Moreover, the administration of an Ado receptor agonist in diabetic rats caused a decrease in the plasma glucose concentration and a decrease in medullary and cortical hydrogen peroxide production, which was associated with a proportional increase in GPx activity, illustrating that the activation of Ado receptors may improve renal antioxidant capacity and glucose metabolism in diabetic rats. A review delineated a central role of purines, their receptors, and enzymes in diabetes by demonstrating that the manipulation of the purinergic axis at different levels can prevent or exacerbate the development and evolution of both type 1 and type 2 diabetes [111].

In the same line, a study with obese rats showed a decrease in GPx and GST activities, besides a decrease in GSH levels in hepatic and renal tissues. The decrease in these enzymatic activities may be due to their rapid consumption and exhaustion of stored GSH levels in fighting RS generated during the development of obesity, which possibly contributes to the progression of obesity-related problems [112].

Purinergic signaling can be exploited in the development of novel therapeutic approaches to treat obesity. Hall et al. [113] showed that ATP could mediate the long-term effects of leptin on blood pressure involved in obesity and hypertension, and high concentrations have been reported to induce inflammatory responses and insulin resistance generation in rat adipocytes [114].

Regarding other metabolic diseases, in a study with an animal model of thyroid disorders, GST and GPx activities and GST protein expression in red blood cells of hyperthyroid and hypothyroid rats were shown to be increased [115]. On the other hand, Baldissarelli et al. [116] demonstrated that the GST activity was decreased in patients with post-thyroidectomy hypothyroidism, probably to preserve high levels of GSH, which can be used by other reactions in the body, such as the neutralization of hydroxyl radicals. Furthermore, the authors also showed an increase in the activity and expression of NTPDase (CD39) and an increase in 5′-nucleotidase and ADA activities, besides a lower concentration of Ado in hypothyroid patients, which was positively correlated with RS levels.

The role of GSH in cancer has also been demonstrated since the decrease in the activity of antioxidant enzymes, such as GPx, and the increase in the levels of damaged DNA bases due to oxidative damage may lead to the formation of free radicals which could induce the appearance of malignant cells [117, 118]. Moreover, Li et al. [119] showed that GR inhibition generates oxidative stress and suppresses lung metastasis and subcutaneous growth of melanoma in vivo. The tumor microenvironment is characterized by unusually high concentrations of ATP and Ado. Ado is a major determinant of the immunosuppressive tumor milieu. In this sense, preclinical data show that targeting the Ado-generating pathway (CD73) or adenosinergic receptors (A2A) relieves immunosuppression and potently inhibits tumor growth [120, 121]. In this context, patients with lung cancer showed a decrease in ADA activity and an increase in A1 receptor expression in lymphocytes, which may contribute to Ado pro-tumor effects by promoting a profile of cytokine levels that favors tumor progression [122].

Oxidative stress, which is implicated in the pathophysiology of neurodegenerative diseases, also affects brain astrocytes. P2Y receptors, largely expressed in the CNS, are proposed to have a cytoprotective action. In the work of Förster and Reiser [123], the potential involvement of P2Y receptors in the antioxidant protection against hydrogen peroxide-induced toxicity in rat brain astrocytes was investigated. Cells were incubated with the wide range P2Y receptor agonist adenosine 5′-(3-thiotriphosphate) (ATPγS) and the particular P2Y1 receptor agonist 2-methylthio-ADP (2MeSADP), and findings showed that levels of GSH were augmented in the presence of both agonists. Moreover, the expression of genes involved in GSH metabolism also relied on the increase of intracellular Ca^{2+} mediated by the P2Y receptor. Taken together, the authors suggest the participation of P2Y receptors in the cytoprotection of astrocytes in the event of oxidative stress.

In the case of neurodegenerative diseases, studies addressing the relationship between glutathione antioxidant enzymes and purinergic signaling have also been performed. Recently, the effect of intracerebroventricular injection of streptozotocin (ICV-STZ), a model of sporadic dementia of the Alzheimer's type, and administration of berberine (BRB) on GSH levels and GST activity was investigated in the cerebral cortex and hippocampus of rats [124]. Both, in the cerebral cortex and hippocampus, the STZ-induced Alzheimer's model significantly decreased GSH levels and GST activity; however, treatment with BRB at the doses of 50 and 100 mg/kg was able to prevent these alterations induced by STZ in rats. Moreover, BRB at both doses also prevented the reduction in NTPDase, 5′-nucleotidase (EC-5′-Nt), and ADA activities in synaptosomes of the cerebral cortex and hippocampus. In this sense, the authors suggested that BRB could have a neuroprotective activity against oxidative stress and purinergic system damage in STZ-induced Alzheimer's model in rats.

Disease	Glutathione system	Reference	Purinergic system	Reference
Acute myocardial infarction	↑ GPx in whole blood	[125]	↑ NTPDase activity in platelets	[99]
	↓ GPx in serum	[126]	↑ ATP, ADP, and AMP hydrolysis ↑ ADA activity in platelets	[134]
	↓ GR in serum	[127]		
Metabolic syndrome (MetS)	↓ GPx and GST activities in liver	[100]	↑ NTPDase, 5'-NT, and NPP activities in platelets ↓ ADA activity in platelets	[102]
	↓ GPx and GST activities in the heart ↓ GSH levels in the heart	[101]	↑ ATP and ADP hydrolysis in lymphocytes ↓ ADA activity in lymphocytes	[103]
Diabetes	↓ GPx and GST activities in liver	[128]	↑ NTPDase, E-NPP, 5'-NT, and ADA activities in platelets	[135]
	↓ GST activity in liver	[129]	↓ NTPDase activity in the cerebral cortex ↓ A1R and ↑A2R in the cerebral cortex	[110]
			↑ ATP and ADP hydrolysis and ADA activity in lymphocytes ↑ NTPDase and ADA activities in platelets ↓ ATP and ↑ ADP and AMP hydrolysis in serum	[109]
Obesity	↓ GPx and GST activities in hepatic and renal tissues ↓ GSH levels in hepatic and renal tissues	[112]	↓ ATP, ADP, and AMP hydrolysis in serum	[136]
			↑ ADA activity in saliva	[137]
Hypothyroidism	↓ GPx activity in serum ↑ T-SH and NPSH concentrations in platelets	[116]	↑ NTPDase, 5'-NT, and ADA activities in platelets ↑ CD39 expression ↓ Ado levels in serum	[116]
	↑ T-SH and NPSH concentrations in serum	[130]	↓ 5'-NT activity in platelets ↑ NPP activity in platelets ↑ AMP and inosine levels in serum	[138]
			↑ CD73 in lymphocytes	[130]

Disease	Glutathione system	Reference	Purinergic system	Reference
Hyperthyroidism	↓ GPx activity in the hippocampus	[131]	↓ NTPDase and 5'-NT activities in platelets ↑ ADA activity in platelets ↑ ATP, ADP, AMP, and inosine in serum ↓ Ado levels in serum	[138]
	↑ GPx activity in the hippocampus	[132]		
Hypertension	↑ GPx activity in mesenteric arteries	[107]	↑ NTPDase, 5'-NT, and ADA activities in platelets	[105]
	↓ GST activity in the kidney	[133]	↑ NTPDase1, NTPDase3, and CD73 expression and activity in the cortex ↑ A2A expression in hippocampus and cortex	[106]
Alzheimer's disease	↓ GSH and GST activity in cortex and hippocampus	[124]	↓ NTPDase, 5'-NT, and ADA activities in cortex and hippocampus	[124]

Table 2.
Changes in glutathione and purinergic systems during diseases.

5. Conclusions

In summary, purinergic and immune systems, comprised mainly of receptors, signaling molecules, and also enzymes, play a key role in many pathologies and regulate the functions especially of the immune system. Besides, these two complex systems closely interact and may modulate GSH levels as well as the expression and activity of glutathione-dependent antioxidant enzymes in both scenarios, health and disease. Future studies will possibly provide more details into the mechanisms underlying the regulation of these enzymes and help to expand the current knowledge.

Acknowledgements

The authors would like to thank the Brazilian agencies Coordenação de Aperfeiçoamento de Pessoal de Nível Superior (CAPES) and Conselho Nacional de Desenvolvimento Científico e Tecnológico (CNPq) for grants and fellowships.

Conflict of interest

The authors declare no conflict of interest.

Author details

Charles Elias Assmann[1], Naiara Stefanello[2], Nathieli Bianchin Bottari[1], Jucimara Baldissarelli[2], Maria Rosa Chitolina Schetinger[1], Vera Maria Melchiors Morsch[1] and Margarete Dulce Bagatini[3]*

1 Federal University of Santa Maria, Santa Maria, Rio Grande do Sul, Brazil

2 Federal University of Pelotas, Pelotas, Rio Grande do Sul, Brazil

3 Federal University of Fronteira Sul, Chapecó, Santa Catarina, Brazil

*Address all correspondence to: margaretebagatini@yahoo.com.br

References

[1] Ballatori N, Krance SM, Notenboom S, Shi S, Tieu K, Hammond CL. Glutathione dysregulation and the etiology and progression of human diseases. Biological Chemistry. 2009;**390**(3): 191-214. DOI: 10.1515/BC.2009.033

[2] Lushchak VI. Glutathione homeostasis and functions: Potential targets for medical interventions. Journal of Amino Acids. 2012;**2012**:1-26. DOI: 10.1155/2012/736837

[3] Thompson JA, Franklin CC. Enhanced glutathione biosynthetic capacity promotes resistance to AS^{3+}-induced apoptosis. Toxicology Letters. 2010;**193**(1):33-40. DOI: 10.1016/j.toxlet.2009.12.004

[4] Cekic C, Linden J. Purinergic regulation of the immune system. Nature Reviews Immunology. 2016;**16**(3):177-192. DOI: 10.1038/nri.2016.4

[5] Chaplin DD. Overview of the immune response. Journal of Allergy and Clinical Immunology. 2010;**125**(2):3-23. DOI: 10.1016/j.jaci.2009.12.980

[6] Cumpstey A, Feelisch M. Free radicals in inflammation. In: Cavaillon J-M, Singer M, editors. Inflammation: From Molecular and Cellular Mechanisms to the Clinic. Weinheim, Germany: Wiley; 2017. p. 695-726. DOI: 10.1002/9783527692156.ch27

[7] Schetinger MRC, Morsch VM, Bonan CD, Wyse ATS. NTPDase and 5′-nucleotidase activities in physiological and disease conditions: New perspectives for human health. BioFactors. 2007;**31**:77-98. DOI: 10.1002/biof.5520310205

[8] Junger WG. Immune cell regulation by autocrine purinergic signalling. Nature Reviews Immunology. 2011;**11**(3):201-212. DOI: 10.1038/nri2938

[9] Burnstock G. Purinergic signalling: Therapeutic developments. Frontiers in Pharmacology. 2017;**8**:1-55. DOI: 10.3389/fphar.2017.00661

[10] Burnstock G. Purinergic nerves. Pharmacological Reviews. 1972;**24**(3):509-581. Available from: https://pubmed.ncbi.nlm.nih.gov/4404211

[11] Burnstock G. Purinergic signalling: Past, present and future. Brazilian Journal of Medical and Biological Research. 2009;**42**(1):3-8. DOI: 10.1590/S0100-879X2008005000037

[12] Purves D, Augustine GJ, Fitzpatrick D, Katz LC, LaMantia A-S, McNamara JO, et al., editors. Neuroscience. 2nd ed. Sunderland (MA): Sinauer Associates; 2001. Available from: https://www.ncbi.nlm.nih.gov/books/NBK10799

[13] Rudolph FB. The biochemistry and physiology of nucleotides. Journal of Nutrition. 1994;**124**(Suppl 1):124S-127S. DOI: 10.1093/jn/124.suppl_1.124S

[14] Seifert R. cCMP and cUMP: Emerging second messengers. Trends in Biochemical Sciences. 2015;**40**(1):8-15. DOI: 10.1016/j.tibs.2014.10.008

[15] Desch M, Schinner E, Kees F, Hofmann F, Seifert R, Schlossmann J. Cyclic cytidine 3′,5′-monophosphate (cCMP) signals via cGMP kinase I. FEBS Letters. 2010;**584**(18):3979-3984. DOI: 10.1016/j.febslet.2010.07.059

[16] Cansev M. Uridine and cytidine in the brain: Their transport and utilization. Brain Research Reviews. 2006;**52**(2):389-397. DOI: 10.1016/j.brainresrev.2006.05.001

[17] O'Dwyer PJ, King SA, Hoth DF, Leyland-Jones B. Role of thymidine in biochemical modulation: A review. Cancer Research. 1987;**47**(15):3911-3919. Available from: https://pubmed.ncbi.nlm.nih.gov/3300957

[18] Hess JR, Greenberg NA. The role of nucleotides in the immune and gastrointestinal systems: Potential clinical applications. Nutrition in Clinical Practice. 2012;**27**(2):281-294. DOI: 10.1177/0884533611434933

[19] Nikiforov M, Bianchi-Smiraglia A. GTP metabolism regulates cancer cell invasion. FASEB Journal. 2015; 29, No. 1_supplement. Available from: https://www.fasebj.org/doi/abs/10.1096/fasebj.29.1_supplement.725.12

[20] Wawrzyniak JA, Bianchi-Smiraglia A, Bshara W, Mannava S, Ackroyd J, Bagati A, et al. A purine nucleotide biosynthesis enzyme guanosine monophosphate reductase is a suppressor of melanoma invasion. Cell Reports. 2013;**5**(2):493-507. DOI: 10.1016/j.celrep.2013.09.015

[21] Blokland A, Schreiber R, Prickaerts J. Improving memory: A role for phosphodiesterases. Current Pharmaceutical Design. 2006;**12**(20):2511-2523. DOI: 10.2174/138161206777698855

[22] Kukreja RC, Salloum FN, Das A. Cyclic guanosine monophosphate signaling and phosphodiesterase-5 inhibitors in cardioprotection. Journal of the American College of Cardiology. 2012;**59**(22):1921-1927. DOI: 10.1016/j.jacc.2011.09.086

[23] Di Virgilio F, Vuerich M. Purinergic signaling in the immune system. Autonomic Neuroscience: Basic and Clinical. 2015;**191**:117-123. DOI: 10.1016/j.autneu.2015.04.011

[24] Burnstock G. Purinergic signaling in the cardiovascular system. Circulation Research. 2017;**120**(1):207-228. DOI: 10.1161/CIRCRESAHA.116.309726

[25] Zimmermann H. Extracellular metabolism of ATP and other nucleotides. Naunyn-Schmiedeberg's Archives of Pharmacology. 2000;**362**(4-5):299-309. DOI: 10.1007/s002100000309

[26] Burnstock G. Purinergic signalling: Pathophysiology and therapeutic potential. Keio Journal of Medicine. 2013;**62**(3):63-73. DOI: 10.2302/kjm.2013-0003-re

[27] Di Virgilio F. Purines, purinergic receptors, and cancer. Cancer Research. 2012;**72**(21):5441-5447. DOI: 10.1158/0008-5472.CAN-12-1600

[28] Burnstock G, Novak I. Purinergic signalling and diabetes. Purinergic Signalling. 2013;**9**(3):307-324. DOI: 10.1007/s11302-013-9359-2

[29] Moura PR, Vidal FAP. Signal transduction: A review about G protein. [Portuguese]. Scientia Medica. 2011;**21**(1):31-36. Available from: http://revistaseletronicas.pucrs.br/ojs/index.php/scientiamedica/article/viewFile/7577/5940

[30] Oka Y, Saraiva LR, Kwan YY, Korsching SI. The fifth class of Gα proteins. Proceedings of the National Academy of Sciences of the USA. 2009;**106**(5):1484-1489. DOI: 10.1073/pnas.0809420106

[31] Hattori M, Gouaux E. Molecular mechanism of ATP binding and ion channel activation in P2X receptors. Nature. 2012;**485**(7397):207-212. DOI: 10.1038/nature11010

[32] Fredholm BB, Abbracchio MP, Burnstock G, Daly JW, Harden TK, Jacobson KA, et al. Nomenclature and classification of purinoceptors. Pharmacological Reviews. 1994;**46**(2):143-156. Available

from: https://pubmed.ncbi.nlm.nih.gov/7938164

[33] Burnstock G, Verkhratsky A. Receptors for purines and pyrimidines. In: Purinergic Signalling and the Nervous System. Springer, Berlin, Heidelberg; 2012; p. 119-244. DOI: 10.1007/978-3-642-28863-0

[34] Antonioli L, Csóka B, Fornai M, Colucci R, Kókai E, Blandizzi C, et al. Adenosine and inflammation: What's new on the horizon? Drug Discovery Today. 2014;**19**(8):1051-1068. DOI: 10.1016/j.drudis.2014.02.010

[35] Ramkumar V, Jhaveri KA, Xie X, Jajoo S, Toth LA. Nuclear factor κB and adenosine receptors: Biochemical and behavioral profiling. Current Neuropharmacology. 2011;**9**(2):342-349. DOI: 10.2174/157015911795596559

[36] Nakav S, Chaimovitz C, Sufaro Y, Lewis EC, Shaked G, Czeiger D, et al. Anti-inflammatory preconditioning by agonists of adenosine A1 receptor. PLoS One. 2008;**3**(5):e2107. DOI: 10.1371/journal.pone.0002107

[37] Antonioli L, Blandizzi C, Pacher P, Haskó G. Immunity, inflammation and cancer: A leading role for adenosine. Nature Reviews Cancer. 2013;**13**(12):842-857. DOI: 10.1038/nrc3613

[38] Dunwiddie TV, Diao L, Kim HO, Jiang JL, Jacobson KA. Activation of hippocampal adenosine A3 receptors produces a desensitization of A1 receptor-mediated responses in rat hippocampus. Journal of Neuroscience. 1997;**17**(2):607-614. DOI: 10.1523/JNEUROSCI.17-02-00607.1997

[39] Coutinho-Silva R, Robson T, Beales PE, Burnstock G. Changes in expression of P2X7 receptors in NOD mouse pancreas during the development of diabetes. Autoimmunity. 2007;**40**(2):108-116. DOI: 10.1080/08916930601118841

[40] Burnstock G, Vaughn B, Robson SC. Purinergic signalling in the liver in health and disease. Purinergic Signalling. 2014;**10**(1):51-70. DOI: 10.1007/s11302-013-9398-8

[41] Delekate A, Füchtemeier M, Schumacher T, Ulbrich C, Foddis M, Petzold GC. Metabotropic P2Y1 receptor signalling mediates astrocytic hyperactivity in vivo in an Alzheimer's disease mouse model. Nature Communications. 2014;**5**:5422. DOI: 10.1038/ncomms6422

[42] Hechler B, Gachet C. Purinergic receptors in thrombosis and inflammation. Arteriosclerosis, Thrombosis, and Vascular Biology. 2015;**35**(11):2307-2315. DOI: 10.1161/ATVBAHA.115.303395

[43] Gold M, El Khoury J. β-Amyloid, microglia, and the inflammasome in Alzheimer's disease. Seminars in Immunopathology. 2015;**37**(6):607-611. DOI: 10.1007/s00281-015-0518-0

[44] Eltzschig HK, Sitkovsky MV, Robson SC. Purinergic signaling during inflammation. The New England Journal of Medicine. 2012;**367**(24):2322-2333. DOI: 10.1056/NEJMra1205750

[45] Idzko M, Ferrari D, Riegel A-K, Eltzschig HK. Extracellular nucleotide and nucleoside signaling in vascular and blood disease. Blood. 2014;**124**(7):1029-1037. DOI: 10.1182/blood-2013-09-402560

[46] Solini A, Usuelli V, Fiorina P. The dark side of extracellular ATP in kidney diseases. Journal of the American Society of Nephrology. 2014;**26**(5):1007-1016. DOI: 10.1681/ASN.2014070721

[47] Glas R, Sauter NS, Schulthess FT, Shu L, Oberholzer J, Maedler K. Purinergic P2X7 receptors regulate

secretion of interleukin-1 receptor antagonist and beta cell function and survival. Diabetologia. 2009;**52**(8):1579-1588. DOI: 10.1007/s00125-009-1349-0

[48] Dinarello CA. Interleukin-1 in the pathogenesis and treatment of inflammatory diseases. Blood. 2011;**117**(14):3720-3732. DOI: 10.1182/blood-2010-07-273417

[49] Ben-Sasson SZ, Hu-Li J, Quiel J, Cauchetaux S, Ratner M, Shapira I, et al. IL-1 acts directly on CD4 T cells to enhance their antigen-driven expansion and differentiation. Proceedings of the National Academy of Sciences of the USA. 2009;**106**(17):7119-7124. DOI: 10.1073/pnas.0902745106

[50] Mackenzie NCW, Huesa C, Rutsch F, MacRae VE. New insights into NPP1 function: Lessons from clinical and animal studies. Bone. 2012;**51**(5):961-968. DOI: 10.1016/j.bone.2012.07.014

[51] Yegutkin GG. Enzymes involved in metabolism of extracellular nucleotides and nucleosides: Functional implications and measurement of activities. Critical Reviews in Biochemistry and Molecular Biology. 2014;**49**(6):473-497. DOI: 10.3109/10409238.2014.953627

[52] Moolenaar WH, Perrakis A. Insights into autotaxin: How to produce and present a lipid mediator. Nature Reviews Molecular Cell Biology. 2011;**12**(10):674-679. DOI: 10.1038/nrm3188

[53] Zimmermann H, Zebisch M, Sträter N. Cellular function and molecular structure of ecto-nucleotidases. Purinergic Signalling. 2012;**8**(3):437-502. DOI: 10.1007/s11302-012-9309-4

[54] Robson SC, Sévigny J, Zimmermann H. The E-NTPDase family of ectonucleotidases: Structure function relationships and pathophysiological significance. Purinergic Signalling. 2006;**2**(2):409-430. DOI: 10.1007/s11302-006-9003-5

[55] Heine P, Braun N, Heilbronn A, Zimmermann H. Functional characterization of rat ecto-ATPase and ecto-ATP diphosphohydrolase after heterologous expression in CHO cells. European Journal of Biochemistry. 1999;**262**(1):102-107. DOI: 10.1046/j.1432-1327.1999.00347.x

[56] Knowles AF, Li C. Molecular cloning and characterization of expressed human ecto-nucleoside triphosphate diphosphohydrolase 8 (E-NTPDase 8) and its soluble extracellular domain. Biochemistry. 2006;**45**(23):7323-7333. DOI: 10.1021/bi052268e

[57] Wang TF, Guidotti G. Golgi localization and functional expression of human uridine diphosphatase. Journal of Biological Chemistry. 1998;**273**(18):11392-11399. DOI: 10.1074/jbc.273.18.11392

[58] Robson SC, Wu Y, Sun X, Knosalla C, Dwyer K, Enjyoji K. Ectonucleotidases of CD39 family modulate vascular inflammation and thrombosis in transplantation. Seminars in Thrombosis and Hemostasis. 2005;**31**(2):217-233. DOI: 10.1055/s-2005-869527

[59] Dwyer KM, Deaglio S, Gao W, Friedman D, Strom TB, Robson SC. CD39 and control of cellular immune responses. Purinergic Signalling. 2007;**3**(1-2):171-180. DOI: 10.1007/s11302-006-9050-y

[60] Langer D, Hammer K, Koszalka P, Schrader J, Robson S, Zimmermann H. Distribution of ectonucleotidases in the rodent brain revisited. Cell and Tissue Research. 2008;**334**(2):199-217. DOI: 10.1007/s00441-008-0681-x

[61] Lavoie EG, Gulbransen BD, Martín-Satué M, Aliagas E, Sharkey KA, Sévigny J. Ectonucleotidases in the digestive system: Focus on NTPDase3 localization. American Journal of Physiology: Gastrointestinal and Liver Physiology. 2011;**300**(4):G608-G620. DOI: 10.1152/ajpgi.00207.2010

[62] Bigonnesse F, Lévesque SA, Kukulski F, Lecka J, Robson SC, Fernandes MJG, et al. Cloning and characterization of mouse nucleoside triphosphate diphosphohydrolase-8. Biochemistry. 2004;**43**(18):5511-5519. DOI: 10.1021/bi0362222

[63] Fausther M, Lecka J, Kukulski F, Lévesque SA, Pelletier J, Zimmermann H, et al. Cloning, purification, and identification of the liver canalicular ecto-ATPase as NTPDase8. American Journal of Physiology: Gastrointestinal and Liver Physiology. 2007;**292**(3):G785-G795. DOI: 10.1152/ajpgi.00293.2006

[64] Biederbick A, Kosan C, Kunz J, Elsässer HP. First apyrase splice variants have different enzymatic properties. Journal of Biological Chemistry. 2000;**275**(25):19018-19024. DOI: 10.1074/jbc.M001245200

[65] Zimmermann H. Ectonucleotidases in the nervous system. In: Chadwick DJ, Goode J, editors. Purinergic Signalling in Neuron-Glia Interactions: Novartis Foundation Symposium 276. Weinheim, Germany: Novartis Foundation; 2006; p. 113-130. DOI: 10.1002/9780470032244.ch10

[66] Sträter N. Ecto-5′-nucleotidase: Structure function relationships. Purinergic Signalling. 2006;**2**(2):343-350. DOI: 10.1007/s11302-006-9000-8

[67] Yegutkin GG. Nucleotide- and nucleoside-converting ectoenzymes: Important modulators of purinergic signalling cascade. Biochimica et Biophysica Acta—Molecular Cell Research. 2008;**1783**(5):673-694. DOI: 10.1016/j.bbamcr.2008.01.024

[68] Colgan SP, Eltzschig HK, Eckle T, Thompson LF. Physiological roles for ecto-5′-nucleotidase (CD73). Purinergic Signalling. 2006;**2**(2):351-360. DOI: 10.1007/s11302-005-5302-5

[69] Dunwiddie TV, Masino SA. The role and regulation of adenosine in the central nervous system. Annual Review of Neuroscience. 2001;**24**:31-55. DOI: 10.1146/annurev.neuro.24.1.31

[70] Moriwaki Y, Yamamoto T, Higashino K. Enzymes involved in purine metabolism—A review of histochemical localization and functional implications. Histology and Histopathology. 1999;**14**(4):1321-1340. DOI: 10.14670/HH-14.1321

[71] Zavialov AV, Yu X, Spillmann D, Lauvau G, Zavialov AV. Structural basis for the growth factor activity of human adenosine deaminase ADA2. Journal of Biological Chemistry. 2010;**285**(16):12367-12377. DOI: 10.1074/jbc.M109.083527

[72] Kaljas Y, Liu C, Skaldin M, Wu C, Zhou Q, Lu Y, et al. Human adenosine deaminases ADA1 and ADA2 bind to different subsets of immune cells. Cellular and Molecular Life Sciences. 2017;**74**(3):555-570. DOI: 10.1007/s00018-016-2357-0

[73] Ravichandran KS. Find-me and eat-me signals in apoptotic cell clearance: Progress and conundrums. Journal of Experimental Medicine. 2010;**207**(9):1807-1817. DOI: 10.1084/jem.20101157

[74] Martinon F, Tschopp J. Inflammatory caspases: Linking an intracellular innate immune system to autoinflammatory diseases. Cell. 2004;**117**(5):561-574. DOI: 10.1016/j.cell.2004.05.004

[75] Martinon F, Mayor A, Tschopp J. The inflammasomes: Guardians of the body. Annual Review in Immunology. 2009;**27**:229-265. DOI: 10.1146/annurev.immunol.021908.132715

[76] Di Virgilio F. Liaisons dangereuses: P2X(7) and the inflammasome. Trends in Pharmacological Sciences. 2007;**28**(9):465-472. DOI: 10.1016/j.tips.2007.07.002

[77] Gombault A, Baron L, Couillin I. ATP release and purinergic signaling in NLRP3 inflammasome activation. Frontiers in Immunology. 2013;**3**:1-7. DOI: 10.3389/fimmu.2012.00414

[78] Elliott MR, Chekeni FB, Trampont PC, Lazarowski ER, Kadl A, Walk SF, et al. Nucleotides released by apoptotic cells act as a find me signal to promote phagocytic clearance. Nature. 2009;**461**(7261):282-286. DOI: 10.1038/nature08296

[79] Eltzschig HK, Eckle T, Mager A, Küper N, Karcher C, Weissmüller T, et al. ATP release from activated neutrophils occurs via connexin 43 and modulates adenosine-dependent endothelial cell function. Circulation Research. 2006;**99**(10):1100-1108. DOI: 10.1161/01.RES.0000250174.31269.70

[80] Chen Y, Corriden R, Inoue Y, Yip L, Hashiguchi N, Zinkernagel A, et al. ATP release guides neutrophil chemotaxis via P2Y2 and A3 receptors. Science. 2006;**314**(5806):1792-1795. DOI: 10.1126/science.1132559

[81] Kälvegren H, Fridfeldt J, Bengtsson T. The role of plasma adenosine deaminase in chemoattractant-stimulated oxygen radical production in neutrophils. European Journal of Cell Biology. 2010;**89**(6):462-467. DOI: 10.1016/j.ejcb.2009.12.004

[82] Oury C, Lecut C, Hego A, Wéra O, Delierneux C. Purinergic control of inflammation and thrombosis: Role of P2X1 receptors. Computational and Structural Biotechnology Journal. 2015;**13**:106-110. DOI: 10.1016/j.csbj.2014.11.008

[83] Vaughan KR, Stokes L, Prince LR, Marriott HM, Meis S, Kassack MU, et al. Inhibition of neutrophil apoptosis by ATP is mediated by the P2Y11 receptor. Journal of Immunology. 2007;**179**(12):8544-8553. DOI: 10.4049/jimmunol.179.12.8544

[84] Wewers MD, Sarkar A. P2X7 receptor and macrophage function. Purinergic Signalling. 2009;**5**(2):189-195. DOI: 10.1007/s11302-009-9131-9

[85] Lenertz LY, Gavala ML, Hill LM, Bertics PJ. Cell signaling via the P2X7 nucleotide receptor: Linkage to ROS production, gene transcription, and receptor trafficking. Purinergic Signal. 2009;**5**(2):175-187. DOI: 10.1007/s11302-009-9133-7

[86] Chen B-C, Lin W-W. Pyrimidinoceptor potentiation of macrophage PGE2 release involved in the induction of nitric oxide synthase. British Journal of Pharmacology. 2000;**130**(4):777-786. DOI: 10.1038/sj.bjp.0703375

[87] Di Virgilio F. Purinergic mechanism in the immune system: A signal of danger for dendritic cells. Purinergic Signal. 2005;**1**(3):205-209. DOI: 10.1007/s11302-005-6312-z

[88] Addi AB, Lefort A, Hua X, Libert F, Communi D, Ledent C, et al. Modulation of murine dendritic cell function by adenine nucleotides and adenosine: Involvement of the A2B receptor. European Journal of Immunology. 2008;**38**(6):1610-1620. DOI: 10.1002/eji.200737781

[89] Novitskiy SV, Ryzhov S, Zaynagetdinov R, Goldstein AE, Huang Y, Tikhomirov OY, et al. Adenosine receptors in regulation

of dendritic cell differentiation and function. Blood. 2008;**112**(5):1822-1831. DOI: 10.1182/blood-2008-02-136325

[90] Wilson JM, Ross WG, Agbai ON, Frazier R, Figler RA, Rieger J, et al. The A2B adenosine receptor impairs the maturation and immunogenicity of dendritic cells. Journal of Immunology. 2009;**182**(8):4616-4623. DOI: 10.4049/jimmunol.0801279

[91] Sullivan GW, Linden J. Role of A2A adenosine receptors in inflammation. Drug Development Research. 1999;**45**(3):103-112. DOI: 10.1189/jlb.0607359

[92] Woehrle T, Yip L, Elkhal A, Sumi Y, Chen Y, Yao Y, et al. Pannexin-1 hemichannel-mediated ATP release together with P2X1 and P2X4 receptors regulate T cell activation at the immune synapse. Blood. 2010;**116**(18):3475-3484. DOI: 10.1182/blood-2010-04-277707

[93] Haskó G, Cronstein B. Regulation of inflammation by adenosine. Frontiers in Immunology. 2013;**4**:1-21. DOI: 10.3389/fimmu.2013.00085

[94] Koscsó B, Csóka B, Selmeczy Z, Himer L, Pacher P, Virág L, et al. Adenosine augments IL-10 production by microglial cells through an A2B adenosine receptor-mediated process. Journal of Immunology. 2012;**188**(1):445-453. DOI: 10.4049/jimmunol.1101224

[95] Haskó G, Szabó C, Németh ZH, Kvetan V, Pastores SM, Vizi ES. Adenosine receptor agonists differentially regulate IL-10, TNF-alpha, and nitric oxide production in RAW 264.7 macrophages and in endotoxemic mice. Journal of Immunology. 1996;**157**(10):4634-4640. Available from: https://pubmed.ncbi.nlm.nih.gov/8906843

[96] Zhang Y, Handy DE, Loscalzo J. Adenosine-dependent induction of glutathione peroxidase 1 in human primary endothelial cells and protection against oxidative stress. Circulation Research. 2005;**96**(8):831-837. DOI: 10.1161/01.RES.0000164401.21929.CF

[97] Leopold JA, Loscalzo J. Oxidative enzymopathies and vascular disease. Arteriosclerosis, Thrombosis, and Vascular Biology. 2005;**25**:1332-1340. DOI: 10.1161/01.ATV.0000163846.51473.09

[98] Husain K, Somani SM. Interaction of exercise and adenosine receptor agonist and antagonist on rat heart antioxidant defense system. Molecular and Cellular Biochemistry. 2005;**270**(1-2):209-214. DOI: 10.1007/s11010-005-5285-0

[99] Bagatini MD, Martins CC, Battisti V, Spanevello RM, Gasparetto D, Rosa CS, et al. Hydrolysis of adenine nucleotides in platelets from patients with acute myocardial infarction. Clinical Biochemistry. 2008;**41**(14-15):1181-1185. DOI: 10.1016/j.clinbiochem.2008.07.008

[100] Réggami Y, Benkhaled A, Boudjelal A, Berredjem H, Amamra A, Benyettou H, et al. *Artemisia herba-alba* aqueous extract improves insulin sensitivity and hepatic steatosis in rodent model of fructose-induced metabolic syndrome. Archives of Physiology and Biochemistry. 2019:1-10. DOI: 10.1080/13813455.2019.1659825. Available from: https://pubmed.ncbi.nlm.nih.gov/31464524/

[101] Pérez-Torres I, Torres-Narváez JC, Guarner-Lans V, Díaz-Díaz E, Perezpeña-Diazconti M, Palacios AR, et al. Myocardial protection from ischemia-reperfusion damage by the antioxidant effect of *Hibiscus sabdariffa* Linnaeus on metabolic syndrome rats. Oxidative Medicine and Cellular Longevity. 2019;**2019**:1724194. DOI: 10.1155/2019/1724194

[102] Martins CC, Bagatini MD, Cardoso AM, Zanini D, Abdalla FH, Baldissarelli J, et al. Regular exercise training reverses ectonucleotidase alterations and reduces hyperaggregation of platelets in metabolic syndrome patients. Clinica Chimica Acta. 2016;**454**:66-71. DOI: 10.1016/j.cca.2015.12.024

[103] Martins CC, Bagatini MD, Cardoso AM, Zanini D, Abdalla FH, Baldissarelli J, et al. Exercise training positively modulates the ectonucleotidase enzymes in lymphocytes of metabolic syndrome patients. International Journal of Sports Medicine. 2016;**37**(12):930-936. DOI: 10.1055/s-0042-114218

[104] Madec S, Rossi C, Chiarugi M, Santini E, Salvati A, Ferrannini E, et al. Adipocyte P2X7 receptors expression: A role in modulating inflammatory response in subjects with metabolic syndrome? Atherosclerosis. 2011;**219**(2):552-558. DOI: 10.1016/j.atherosclerosis.2011.09.012

[105] Cardoso AM, Bagatini MD, Martins CC, Abdalla FH, Zanini D, Schmatz R, et al. Exercise training prevents ecto-nucleotidases alterations in platelets of hypertensive rats. Molecular and Cellular Biochemistry. 2012;**371**(1-2):147-156. DOI: 10.1007/s11010-012-1431-7

[106] Cardoso AM, Manfredi LH, Zanini D, Bagatini MD, Gutierres JM, Carvalho F, et al. Physical exercise prevents memory impairment in an animal model of hypertension through modulation of CD39 and CD73 activities and A2A receptor expression. Journal of Hypertension. 2019;**37**(1):135-143. DOI: 10.1097/HJH.0000000000001845

[107] Sousa T, Pinho D, Morato M, Marques-Lopes J, Fernandes E, Afonso J, et al. Role of superoxide and hydrogen peroxide in hypertension induced by an antagonist of adenosine receptors. European Journal of Pharmacology. 2008;**588**(2-3):267-276. DOI: 10.1016/j.ejphar.2008.04.044

[108] Merighi S, Borea PA, Gessi S. Adenosine receptors and diabetes: Focus on the A2B adenosine receptor subtype. Pharmacological Research. 2015;**99**:229-236. DOI: 10.1016/j.phrs.2015.06.015

[109] Pereira AS, Oliveira LS, Lopes TF, Baldissarelli J, Palma TV, Soares MSP, et al. Effect of gallic acid on purinergic signaling in lymphocytes, platelets, and serum of diabetic rats. Biomedicine & Pharmacotherapy. 2018;**101**:30-36. DOI: 10.1016/j.biopha.2018.02.029

[110] Reichert KP, Schetinger MRC, Gutierres JM, Pelinson LP, Stefanello N, Dalenogare DP, et al. Lingonberry extract provides neuroprotection by regulating the purinergic system and reducing oxidative stress in diabetic rats. Molecular Nutrition & Food Research. 2018;**62**(16):e1800050. DOI: 10.1002/mnfr.201800050

[111] Fotino C, Dal Ben D, Adinolfi E. Emerging roles of purinergic signaling in diabetes. Medicinal Chemistry. 2018;**14**(5):428-438. DOI: 10.2174/1573406414666180226165204

[112] Noeman SA, Hamooda HE, Baalash AA. Biochemical study of oxidative stress markers in the liver, kidney and heart of high fat diet induced obesity in rats. Diabetology & Metabolic Syndrome. 2011;**3**(1):17. DOI: 10.1186/1758-5996-3-17

[113] Hall JE, Hildebrandt DA, Kuo J. Obesity hypertension: Role of leptin and sympathetic nervous system. American Journal of Hypertension. 2001;**14** (6 Pt 2):103S-115S. DOI: 10.1016/s0895-7061(01)02077-5

[114] Yu Z, Jin T. Extracellular high dosages of adenosine triphosphate induce inflammatory response and insulin resistance in rat adipocytes.

Biochemical and Biophysical Research Communications. 2010;**402**(3):455-460. DOI: 10.1016/j.bbrc.2010.10.028

[115] Araujo ASR, Seibel FER, Oliveira UO, Fernandes T, Llesuy S, Kucharski L, et al. Thyroid hormone-induced haemoglobin changes and antioxidant enzymes response in erythrocytes. Cell Biochemistry and Function. 2011;**29**(5):408-413. DOI: 10.1002/cbf.1765

[116] Baldissarelli J, Pillat MM, Schmatz R, Cardoso AM, Abdalla FH, de Oliveira JS, et al. Post-thyroidectomy hypothyroidism increases the expression and activity of ectonucleotidases in platelets: Possible involvement of reactive oxygen species. Platelets. 2018;**29**(8):801-810. DOI: 10.1080/09537104.2017.1361017

[117] Traverso N, Ricciarelli R, Nitti M, Marengo B, Furfaro AL, Pronzato MA, et al. Role of glutathione in cancer progression and chemoresistance. Oxidative Medicine and Cellular Longevity. 2013;**2013**:972913. DOI: 10.1155/2013/972913

[118] Bansal A, Simon MC. Glutathione metabolism in cancer progression and treatment resistance. Journal of Cell Biology. 2018;**217**(7):2291-2298. DOI: 10.1083/jcb.201804161

[119] Li X, Wu J, Zhang X, Chen W. Glutathione reductase-mediated thiol oxidative stress suppresses metastasis of murine melanoma cells. Free Radical Biology and Medicine. 2018;**129**:256-267. DOI: 10.1016/j.freeradbiomed.2018.07.025

[120] Di Virgilio F, Adinolfi E. Extracellular purines, purinergic receptors and tumor growth. Oncogene. 2017;**36**(3):293-303. DOI: 10.1038/onc.2016.206

[121] Di Virgilio F, Sarti AC, Falzoni S, De Marchi E, Adinolfi E. Extracellular ATP and P2 purinergic signalling in the tumour microenvironment. Nature Reviews Cancer. 2018;**18**(10):601-618. DOI: 10.1038/s41568-018-0037-0

[122] Zanini D, Manfredi LH, Pelinson LP, Pimentel VC, Cardoso AM, Gonçalves VCA, et al. ADA activity is decreased in lymphocytes from patients with advanced stage of lung cancer. Medical Oncology. 2019;**36**(9):78. DOI: 10.1007/s12032-019-1301-1

[123] Förster D, Reiser G. Nucleotides protect rat brain astrocytes against hydrogen peroxide toxicity and induce antioxidant defense via P2Y receptors. Neurochemistry International. 2016;**94**:57-66. DOI: 10.1016/j.neuint.2016.02.006

[124] Oliveira JS, Abdalla FH, Dornelles GL, Palma TV, Signor C, Bernardi JS, et al. Neuroprotective effects of berberine on recognition memory impairment, oxidative stress, and damage to the purinergic system in rats submitted to intracerebroventricular injection of streptozotocin. Psychopharmacology. 2019;**236**(2):641-655. DOI: 10.1007/s00213-018-5090-6

[125] Madole MB, Bachewar NP, Aiyar CM. Study of oxidants and antioxidants in patients of acute myocardial infarction. Advanced Biomedical Research. 2015;**4**:241. DOI: 10.4103/2277-9175.168608

[126] Siddiqui AH, Gulati R, Tauheed N, Pervez A. Correlation of waist-to-hip ratio (WHR) and oxidative stress in patients of acute myocardial infarction (AMI). Journal of Clinical and Diagnostic Research. 2014;**8**(1):4-7. DOI: 10.7860/JCDR/2014/6446.3912

[127] Janahmadi Z, Nekooeian AA, Moaref AR, Emamghoreishi M. Oleuropein offers cardioprotection in rats with acute myocardial infarction. Cardiovascular Toxicology. 2015;**15**(1):61-68. DOI: 10.1007/s12012-014-9271-1

[128] Sekiou O, Boumendjel M, Taibi F, Boumendjel A, Messarah M. Mitigating effects of antioxidant properties of *Artemisia herba alba* aqueous extract on hyperlipidemia and oxidative damage in alloxan-induced diabetic rats. Archives of Physiology and Biochemistry. 2019;**125**(2):163-173. DOI: 10.1080/13813455.2018.1443470

[129] Oliveira LS, Thomé GR, Lopes TF, Reichert KP, de Oliveira JS, da Silva PA, et al. Effects of gallic acid on delta-aminolevulinic dehydratase activity and in the biochemical, histological and oxidative stress parameters in the liver and kidney of diabetic rats. Biomedicine & Pharmacotherapy. 2016;**84**:1291-1299. DOI: 10.1016/j.biopha.2016.10.021

[130] Baldissarelli J, Mânica A, Pillat MM, Bagatini MD, Leal DBR, Abdalla FH, et al. Increased cytokines production and oxidative stress are related with purinergic signaling and cell survival in post-thyroidectomy hypothyroidism. Molecular and Cellular Endocrinology. 2020;**499**:110594. DOI: 10.1016/j.mce.2019.110594

[131] Rao G, Verma R, Mukherjee A, Haldar C, Agrawal NK. Melatonin alleviates hyperthyroidism induced oxidative stress and neuronal cell death in hippocampus of aged female golden hamster, *Mesocricetus auratus*. Experimental Gerontology. 2016;**82**:125-130. DOI: 10.1016/j.exger.2016.06.014

[132] Tan B, Babur E, Koşar B, Varol S, Dursun N, Süer C. Age-dependent evaluation of long-term depression responses in hyperthyroid rats: Possible roles of oxidative intracellular redox status. Brain Research. 2019;**1720**:146314. DOI: 10.1016/j.brainres.2019.146314

[133] Javkhedkar AA, Quiroz Y, Rodriguez-Iturbe B, Vaziri ND, Lokhandwala MF, Banday AA. Resveratrol restored Nrf2 function, reduced renal inflammation, and mitigated hypertension in spontaneously hypertensive rats. American Journal of Physiology: Regulatory, Integrative and Comparative Physiology. 2015;**308**(10):R840-R846. DOI: 10.1152/ajpregu.00308.2014

[134] Lavall MC, Bagatini MD, Thomé GR, Bonfanti G, Moretto MB, De Oliveira LZ, et al. Extracellular hydrolysis of adenine nucleotides and nucleoside adenosine is higher in patients with ST elevation than non-ST elevation in acute myocardial infarction. Clinical Laboratory. 2015;**61**(7):761-767. DOI: 10.7754/clin.lab.2014.141136

[135] Schmatz R, Mann TR, Spanevello R, Machado MM, Zanini D, Pimentel VC, et al. Moderate red wine and grape juice consumption modulates the hydrolysis of the adenine nucleotides and decreases platelet aggregation in streptozotocin-induced diabetic rats. Cellular Biochemistry and Biophysics. 2013;**65**(2):129-143. DOI: 10.1007/s12013-012-9407-5

[136] Souza CG, Böhmer AE, Müller AP, Oses JP, Viola GG, Lesczinski DN, et al. Effects of a highly palatable diet on lipid and glucose parameters, nitric oxide, and ectonucleotidases activity. Applied Physiology, Nutrition, and Metabolism. 2010;**35**(5):591-597. DOI: 10.1139/H10-048

[137] Chielle EO, Bonfanti G, De Bona KS, Moresco RN, Moretto MB. Adenosine deaminase, dipeptidyl peptidase-IV activities and lipid peroxidation are increased in the saliva of obese young adult. Clinical Chemistry and Laboratory Medicine. 2015;**53**(7):1041-1047. DOI: 10.1515/cclm-2014-1086

[138] Baldissarelsli J, Santi A, Schmatz R, Martins CC, Zanini D, Reichert KP, et al. Hypothyroidism and hyperthyroidism change ectoenzyme activity in rat platelets. Journal of Cellular Biochemistry. 2018;**119**(7):6249-6257. DOI: 10.1002/jcb.26856

Chapter 2

Subcellular Localization of Glutathione Peroxidase, Change in Glutathione System during Ageing and Effects on Cardiometabolic Risks and Associated Diseases

Théophile Mbemba Fundu, Paulin Mutwale Kapepula, Justin Mboloko Esimo, José Remacle and Nadege Kabamba Ngombe

Abstract

Glutathione peroxidase (GPx) is a selenoprotein with biological properties that allow the detoxification of endogenous or exogenous reactive oxygen species as well as the elimination of xenobiotic compounds in the cells. Due to its isoform activities and pathophysiological functions, GPx holds the status of a redox system (GSH/GSSG) in the glutathione (GSH) system to prevent oxidative damage of cellular constituents. As such, the GPx is the first line of defense against free radicals. Its deficiency causes oxidative stress that not only promotes the oxidation of proteins and deoxyribonucleic acid (DNA) but also leads to insulin resistance, dyslipidemia, inflammation, and metabolic alterations, which expose to high risk for cardiometabolic disorders due to cardiovascular and degenerative diseases especially when associated with aging. This work presents a review of different studies done on the localization of GPx in subcellular organelles, activity changes during cellular aging, their effects on cardiometabolic risks, and associated diseases.

Keywords: aging, antioxidants, cardiometabolic risks, disease, free radicals, glutathione, glutathione peroxidase, selenium, traditional foods

1. Introduction

Cardiometabolic risks (CMR), the main causes of the onset of cardiovascular diseases (CVDs), insulin resistance (IR), dyslipidemia, and systemic inflammation are among the major metabolic alterations caused largely by oxidative stress. The oxidative stress is the result of the imbalance of the antioxidant system in favor of prooxidants, which interferes with the GSH/GSSG system in the antioxidant defense and the regulation of gene expression, the synthesis of DNA and proteins, cell proliferation and apoptosis, and cytokine production and protein glutathionylation, due to alteration of certain cellular functions. In this system, the deficiency of GPx, as a first line of defense against free radicals, stimulates oxidative stress, which promotes the development of chronic noncommunicable diseases (NCDs), such as CVDs, as well as early aging and cancer [1]. Having sufficient knowledge of the GSH system and the regulation and functions of GPx is essential to prevent metabolic alterations and to develop effective strategies for treating these diseases. This chapter reviews (i) the main scientific information on GPxs and the GSH system and their location in subcellular organelles and changes during aging; (ii) the link between oxidative stress, GPxs, and the metabolic syndrome; and (iii) the effects of GPxs in chronic pathogenesis and CVDs in particular and the role of dietary selenium (Se) in GPx activities.

2. Glutathione system

2.1 Structure and functions of the glutathione system

Glutathione, γ-glutamyl-cysteinyl-glycine (GSH), is a ubiquitous intracellular tripeptide present in all mammalian tissues, especially in the liver. This thiol-containing molecule is an important antioxidant in cell compartments with high concentrations in cytosol (1–11 mM), nuclei (3–15 mM), and mitochondria (5–11 mM). GSH represents a significant part of the redox status of thiol mammalian systems [2, 3]. It should be noted that the research on GSH metabolites was done in vivo more than a century before [4].

GSH has several biological functions including the detoxification of electrophiles, the antioxidant defense, the maintenance of the thiol status of proteins, and the modulation of DNA synthesis and the immune system [4]. Additionally, GSH serves as a cysteine reservoir with a proton-donating sulfhydryl function, which allows GSH to act as an antioxidant. In its role as an antioxidant, GSH effectively removes free radicals and other reactive oxygen species through the GPx activity [5], which oxidizes GSH to GSSG, and the action of NADPH-dependent glutathione reductase, which generates GSH [3]. In the presence of GSH, glutathione-S-transferase activity detoxifies xenobiotics and various physiological metabolites to form mercapturates and reactivated glucose-6-phosphate dehydrogenase [6]. With NO, GSH is necessary for the hepatic action of insulin sensitizing agents and plays a crucial role in regulating the redox state of the cell with lipids, glucose, and amino acids. Besides to its antioxidant nature, GSH is involved in the transfer of amino acids by the gamma-glutamyl cycle as well as in the hormonal metabolism of estrogen, leukotrienes, prostaglandins and as a transduction signal for transcription [3]. In the central nervous system (CNS), GSH functions include maintenance of neurotransmitters, membrane protection, detoxification, metabolic regulation, and modulation of signal transduction. The depletion of GSH in the brain is implicated in both Parkinson's disease and neuronal damage after stroke [7].

2.2 Regulation of glutathione metabolism

Glutathione is synthesized in the cytosol of all animal cells by the regulated action of gamma-glutamate cysteine ligase (γ-GCL), glutathione synthetase, glutathione reductase (GSR), and gamma-glutamyl transpeptidase (γ-GGT) and from which it is distributed to the other cellular compartments [8]. The key transcription factors that regulate gene expression are NF-E2-related factor 2 (Nrf2) via the antioxidant response element (ARE), AP-1, and nuclear factor kappa B (NF-κB). The alteration of the GSH concentration affects the dysregulation of cell proliferation and the transcription of detoxification enzymes and apoptosis [9]. Therefore, de novo synthesis of GSH is essential for the adaptive response to oxidative stress.

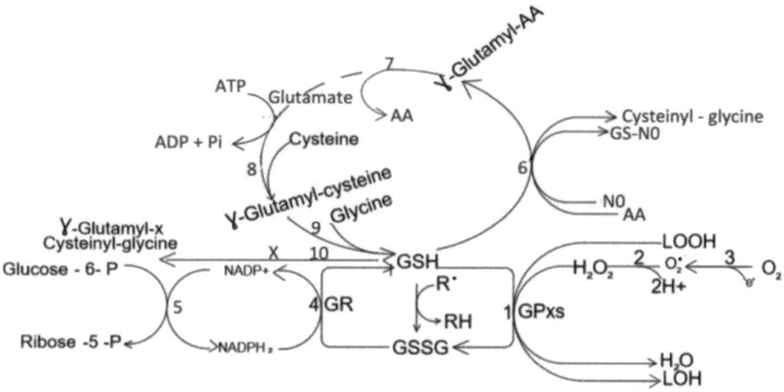

Figure 1.
Glutathione system with synthesis routes and pentose phosphates: the enzymes that catalyze the reactions are (1) GPxs, (2) superoxide dismutase, (3) NADPH oxidase and mitochondrial respiratory complexes, (4) glutathione reductase, (5) gluco-6-phosphate dehydrogenase, (6) γ-glutamyl transpeptidase, (7) γ-glutamyl cyclotransferase, (8) γ-glutamylcysteine synthetase, (9) glutathione synthetase, (10) γ-glutamyl transpeptidase. Abbreviations: AA, amino acid; O_2^-, radical superoxide; H_2O_2: hydrogen peroxide; GS-NO, glutathione-nitric oxide adduct; LOH, alcohol lipid; LOOH, hydroperoxide lipid, R·, radical; RH, nonradical; X, electrophilic xenobiotics.

Intracellular GSH homeostasis is regulated not only by de novo synthesis but also by several factors, including its use and recycling in cells [3]. A disruption of GSH homeostasis could induce oxidative stress and lead to neurodegenerative diseases, including amyotrophic lateral sclerosis, Parkinson's disease, Alzheimer's disease, and dementia, with impaired motor and cognitive functions [7]. There is also increasing evidence that deregulation of GSH synthesis contributes to the pathogenesis of diseases such as diabetes mellitus, pulmonary and hepatic fibrosis, alcoholic liver disease, cholestatic liver injury, endotoxemia, and drug-resistant tumors cells [10]. GSH also modulates cell death whether it is apoptosis or necrosis. In both cases, GSH levels influence the expression/activity of caspases and other important signaling molecules in cell death. The regulation of GSH is well reported by Lu [11], and its role is illustrated in **Figure 1**.

3. GPx, oxidative stress and cardiometabolic risk

GPx is the most powerful biological antioxidant reducer. The GSH/GSSG ratio of GSH, as well as other active redox couples, including NADP/NADPH and FAD/FADH2, regulates and maintains cellular redox status. Under normal conditions, antioxidant systems neutralize ROS. However, when the ROS level is high metabolic alterations of cellular constituents occurs related to oxidative damage to the cells [6, 10]. When there is a prolonged increase in oxygen reactive species (ROS) levels that the existing antioxidant potential cannot eliminate, cell enters in the state of chronic oxidative stress. This leads to insulin resistance (IR), atherogenic dyslipidemia, visceral obesity, and pro-inflammatory and pro-thrombotic status. These are potential factors that increase cardiometabolic risks (CMRSs) and may lead to or accompany some pathologies, such as diabetes mellitus, cardiovascular disease, neurodegenerative, cancer, and aging diseases [12].

The GSH redox cycle is a major source of protection against mild oxidative stress with the GPx, antioxidant enzyme that oxidizes GSH to GSSG to protect cells against the proliferation of reactive oxygen species (ROS) or reactive nitrogen species (RNS), sparing them from oxidative damage, while catalase is becoming increasingly important in protecting against severe oxidative stress [3, 6, 10]. The variation in the erythrocyte GSH system without nuclear capacity to restore homeostasis may be an early biomarker of chronic oxidative stress that could be a first step in the development of cardiometabolic complications. If the cells are overwhelmed by the intensity of the oxidative stress, they die by necrosis or apoptosis [13].

It has also been reported that obese women with high GPx activity have an altered cardiometabolic profile, evidenced by insulin resistance predominantly affecting the liver, altered carbohydrate and lipid metabolisms, and a larger wall thickness of blood vessels than those with lower GPx activity. This suggests that GPx blood activity may be a parameter contributing to the identification of subclinical asymptomatic cardiometabolic disorders [14].

4. Subcellular localization of GPx and change of the glutathione system during aging

In view of its role in the regulation of the cellular redox status, GSH has specific vital functions within the intracellular organelles where it is located. GSH is generally in the greatly reduced state in the different cellular compartments. The integrity of cell and subcellular membranes is highly dependent on the presence of GSH and GPx [15]. Decreased GSH levels in some organelles and tissues during aging expose cells to an increased risk of succumbing to stress. Moderate stress increases glutathione levels to protect cells against more severe stress. In the cytoplasm, the oxidized form (GSSG) is usually in the order of at least about 1% of the total. In the nucleus, GSH maintains the redox status of the sulfhydryl groups of the proteins involved in nucleic acid biosynthesis and DNA repair. It is also used in the reduction of ribonucleotides to produce deoxyribonucleotides by ribonucleotide reductase [16]. A significant portion of ER glutathione is oxidized, with a [GSH]/[GSSG] ratio that can reach 3:1. This relatively oxidative thiol-disulfide medium is essential for the oxidative folding of nascent proteins in ER. Mitochondria contain 10–15% of cellular GSH, but being of a very small volume, the local concentration of GSH in these organelles is generally great and 85–90% in the cytosol. Studies have shown that there is a close relationship between the survival of the mitochondrial GSH pool (mGSH) and that of the cells due to the central role of mitochondria in programmed cell death (apoptosis) as well as important involvement of ROS produced at 90% in mitochondria [17]. High levels of ROS and calcium, acting together, can trigger the mechanism of cell death via apoptosis or necrosis. Thus, the decrease in mGSH levels is closely associated with certain pathologies in both humans and animals. Differential centrifugation and isopycnic equilibration in WI-38 fibroblast density gradients allowed for GSH localization in all subcellular fractions, whereas glutathione peroxidase and reductase activities were restricted to cytoplasm and mitochondrial fractions. The evolution of GSH in aging fibroblasts showed a sudden increase in its concentration just before cell death, whereas GPx activity was already decreasing at the beginning of passages, and that of glutathione reductase was constant and reaching a very low level at the end of the cell culture, suggesting that the GSH system was probably involved in cell degeneration associated with aging [18].

Glutathione peroxidase (EC 1.11.1.9 and EC 1.11.1.12) is a superfamily of proteins found in many living organisms. It consists of four subunits generally containing a Se atom incorporated in the form of selenocysteine (Sec), which is recognized today as the twenty-first amino acid, the first major enzyme identified as an intracellular antioxidant. In animal cells, and in particular in human erythrocytes, GPx is the main antioxidant enzyme for detoxification hydrogen peroxide (H_2O_2) [17]. It is dependent on Se; the deficiency of which is associated with the risk of contracting several diseases, notably cancer. GPxs that use GSH to catalyze the reduction of H_2O_2 and lipid peroxides have been identified. Some GPxs are therefore dependent on Se and use GSH as a reducing agent, while others, called TGPx, do not contain Se (NS-GPx) and reduce ROS using thioredoxin, which acts as ROS sensors in various pathways and signal transduction. Catalysis of GPx is essentially following three distinct redox modifications of the Se at the center of the active site, in a triad of selenocysteine, glutamine, and tryptophan, which reduces GSH to GSSG [14, 19].

The study on the evolution of the gene family of GPxs suggests that classes of basal peroxidase glutathione originate from independent evolutionary events such as gene duplication, gene loss, and lateral transfer of genes between invertebrates and vertebrates or plants. This evolution of the family of the GPx gene as a whole has been described by Deponte and Margis et al. [20, 21]. The mammalian GPx family is divided into six clades according to their amino acid sequence, substrate, specificity, and subcellular localization. Other studies have revealed other GPx containing as peroxidic site the residue of cysteine. These include the GPx7 and GPx8, which are isoforms of endoplasmic reticulum sulfhydryl peroxidases.

As a reminder, in humans, eight different isoforms of GPx (GPx1–8), which use GSH to catalyze the reduction of hydrogen peroxide and lipid peroxides, have been identified [19, 20]: GPx1, GPx2, GPx3, GPx4, and GPx6, which contain a selenocysteine residue (SeCys) and the GPx5, GPx7, and GPx8 that do not contain SeCys but contain a Cys residue. Noteworthy, GPx1, GPx2, GPx4, GPx5, GPx6, and GPx7 are tetramers, while GPx3 is monomeric and GPx8 is dimeric [20, 22, 23]. The following paragraphs briefly provide an overview of each of the eight types of GPx.

- Glutathione peroxidase 1 (glutathione: H_2O_2 oxidoreductase, EC 1.11.1.9, GPx1 or cGPx) is abundant in the cytoplasm of almost all mammalian tissues. Its gene is characterized by the Pro198Leu polymorphism and a number of leucine-repeated alanine (A7L) codons, which are associated with the risk of cancer and type 2 diabetes. GPx1 prevents oxidative damage, lipid peroxidation, and protein degradation induced by cytotoxic peroxides. GSH cytosolic and mitochondrial peroxidases only reduce soluble hydroperoxides, such as H_2O_2 and some organic hydroperoxides, such as hydroperoxy fatty acids, cumenehydro peroxide, or t-butyl hydroperoxide. Increased activity of GPx1 inhibits hydroperoxide-induced apoptosis. GPx1 and phospholipid hydroperoxide glutathione peroxidase GPx4 (or PHGPx) are found in most tissues [17, 19]. A study of the subcellular localization of GPx1 variants to appreciate the molecular consequences associated with diseases demonstrated that the primary sequence of GPx1 affects subcellular localization and that the sequence and cell location can be important to understand the impact of GPx1 on human diseases, including cancer [24].

- Glutathione peroxidase 2 (GPx2 or GI-GPx) is extracellular and important as a barrier against the absorption of hydroperoxide in the gastrointestinal tract. GPx2 could be an anti-inflammatory and anticarcinogenic enzyme [25].

- Glutathione peroxidase 3 (GPx3 or pGPx) is excreted from various tissues in contact with body fluids and is particularly abundant in plasma. It reduces hydroperoxides of phospholipids and contributes to extracellular antioxidant status in humans. Low levels of GPx3 increase the risk of cardiovascular events in patients with a trial fibrillation and in the elderly. GPx3 is directed to extracellular compartments [23].

- GPx4 or PHGPx is located in the cytosol and membrane fraction. It reduces more complex lipids such as phosphatidylcholine hydroperoxides, fatty acid hydroperoxides, and cholesterol [26]. GPx4 shares the amino acid motif of selenocysteine, glutamine, and tryptophan (catalytic triad) with other GPxs. Its inactivation causes an accumulation of lipid peroxides, resulting in the death of ferroptotic cells and mutations causing spondylometaphyseal dysplasia [27]. In mice and rats, three distinct GPx4 isoforms, cytosolic GPx4, mitochondrial GPx4 (mGPx4), and nuclear GPx4 (nGPx4), were identified with different functions. Cytosolic GPx4 is essential for embryonic development and cell survival. Both mGPx4 and nGPx4 are involved in spermatogenesis and male fertility. GPx4 has been shown to be a more attractive candidate for silencing lipooxygenases and influencing cytokine signaling [27].

- Glutathione peroxidase (GPx5) does not contain Sec or Se; it is specifically expressed in the epididymis of the male reproductive tract in mammals and is regulated by androgens. It plays a role in the protection of sperm membranes against the harmful effects of lipid peroxidation and/or in preventing the premature reaction of the acrosome [21].

- Glutathione peroxidase 6 (GPx6) is a selenocysteine close to GPx3 whose expression of its gene is limited to embryos and adult olfactory epithelium [19, 21].

- Glutathione peroxidase 7 (GPx7) is an endoplasmic reticulum (ER) monomer containing a Cys redox center (CysGPx). It catalyzes the peroxidase cycle through a Cys mechanism in which GSH and protein disulfide isomerases,are alternative substrates, allowing rapid reactivity with thioredoxin (Trx) or proteins related to most other CysGPx. It protects esophageal epithelia and breast cancer cells from oxidative stress [19].

- Glutathione peroxidase 8 (GPx8) is a resident endoplasmic reticulum (ER) protein that introduces disulfide bonds into nascent proteins via protein disulfide isomerase (PDI); it is a PDI peroxidase that reduces the H_2O_2 content and oxidative stress in emergency rooms [21]. In the presence of peroxide, GPx7 and GPx8 interact by oxidation for the folding of disulfide-forming proteins.

5. Metabolic regulation of glutathione peroxidase

Many studies on the metabolic regulation of GPx have been focused on GPx1, a selenocysteine-dependent enzyme (Sec). The latter is encoded by UGA and directed by the selenocysteine insertion sequence, SECIS element [28, 29], which serves as a platform for the recruitment of elongation factors of selenocysteine-tRNA$^{[Ser]Sec}$ (Sec-tRNA $^{[Ser]\ Sec}$) translation that decodes the UGA codon for the whole family of selenoproteins. The exogenous Se supply controls the enzymatic activity of human GPx1 without affecting the level of GPx1 mRNA, suggesting that the human GPx1 gene is posttranscriptionally regulated by Se [30].

GPx1 is induced by etoposide, topoisomerase II inhibitor, apoptosis inducer, and p53 activator, which positively regulates a promoter element upstream of the GPx1 gene. This transactivation of GPx1 by p53 bonds is the p53 signaling pathway to the antioxidant pathway. In addition, analysis of p53-induced apoptosis in a human colon cancer cell line showed that elevated p53 expression was associated with an elevation of GPx1 [31, 32]. Studies have shown that in the skeletal muscle of severely dyslipidemic transgenic mice and in a pro-oxidative and pro-inflammatory state, GPx1 is hypermethylated, which decreases GPx1 expression and weakens the endogenous antioxidant defense. The chronic physical exercise allowed increasing the expression of GPx1 in connection with a transient hypomethylation of its gene. The epigenetic regulation of the expression of GPx1 is therefore a function of the methylation of its coding gene [32]. As part of this review, we report knowledge on regulatory factors, the link between the regulation of the mRNA and the expression of GPx activity, the relationship between abnormal expression of GPx1 and the ethology of diseases, and finally the roles of GPx in different diseases especially in chronic diseases.

In addition to Se as the main regulator of GPx1 expression, the factors associating the selenocysteine insertion sequence (SECIS), adenosine [33], c-Abl and Arg tyrosine kinase receptors, and epidermal growth factor influence gene expression of GPx1 and affect the functional coordination between GPx1 and other selenoproteins or antioxidant enzymes in various metabolic circumstances. It will also be interesting to know how these regulators affect the functional coordination between GPx1 and other selenoproteins or antioxidant enzymes in various metabolic circumstances [34]. Se is the main regulator of GPx1 expression. Lower levels of Se cause a decrease in GPx activity [35], increasing the damage caused by free radicals, which contribute to aging and mortality in adults over 65 [36, 37]. In the cells, Se deficiency results in a 60% reduction in GPx1 mRNA and a 93% loss of GPx1 activity. The injection of dietary Se allows rapid recovery of a saturable activity of GPx1. Of these facts, GPx1 was used as a biomarker for assessing body status in Se or Se nutritional status requirement; it is also considered a place of storage of Se for the regulation of the expression of selenoprotein [38, 39]. Se reduces the incidence of aberrant preneoplastic colon cancer and crypt foci in animal models [40]. He has also been involved in the possible chemoprevention of certain cancers [41]. Using a mouse TGFα/c-Myc model of cancer, Novoselov et al. suggested that selenoproteins and Se compounds contributed to the inhibition of liver carcinogenesis. Although GPx1 is unlikely to be the only selenoprotein involved, these results have suggested the involvement of GPx1 in chemotherapy prevention conferred by the food Se [42].

The SECIS selenocysteine (sec) insertion sequence represents mRNA and serves as a platform for the recruitment of Sec -ARNt translation stretching factors that decode the UGA codon for the incorporation of Se into selenoproteins [43]. The SECIS association factors regulate the expression of selenoprotein by displacement of SECIS-binding protein 2 (SBP2), which specifically binds SECIS to dry EF elongation factor with specificity for selenocysteyl-tRNA (Sec-tRNA$^{[Ser]\ Sec}$). TRNA (Ser) SECIS is aminoacylated with serine which is then converted to intracellular Sec [44] from cytoplasm to nucleus in case of exposure to ROS or depending on Se status that modulates Sec-tRNA$^{[Ser]\ Sec}$ with methylation of Sec in position [45], which modifies the secondary and tertiary structure of Sec-tRNA $^{[Ser]\ Sec}$. It is for this reason that the expression of GPx1 and GPx3 is highly reactive to the deficiency in Se. The basic mechanisms of the synthesis and insertion of Sec in proteins, their characterization, the molecular and physiological functions of selenoproteins, and their roles in human health were reviewed by Vyacheslav [46].

Adenosine is a powerful and independent regulator of GPx expression; it attenuates damaging effects of ROS in the cells and improves the stability of the mRNA [47]. The non-receptor c-Abl and Arg tyrosine kinases represent another Se-independent regulator for GPx1 expression. They are activated in the response to ROS and involved in the apoptotic response to oxidative stress. C-Abl and Arg combine, and their interaction is regulated by the intracellular level of oxidants. GPx1 functions as a substrate for c-Abl- and Arg-mediated phosphorylations in Tyr-96, which induces its activity. Loss of GPx1 regulation by c-Abl and Arg increases the susceptibility to ROS induced by apoptosis [48]. NF-κB is a transcription factor involved in the regulation of cellular responses to a variety of environmental stressors [49]. Recent evidence has suggested that GPx1 and c-Src tyrosine kinases participate in the phosphorylation of IκBα which, in response to hypoxia, leads to the activation of NF-κB elevated in hydrogen peroxide-treated embryonic fibroblasts [50]. GPxs modulate the activation of NF-κB inhibitors by cyclooxygenases and lipooxygenases, the activation which depends on hydroperoxide. It also neutralizes hydroperoxide effects, such as cytokine signal and apoptosis, and also has an important role in the human immunodeficiency virus (HIV) infection. The mitogen-activated protein kinase p38 (MAPK) and c-Jun N-terminal kinase (JNK) transmit essential information in ROS-induced apoptosis [51]. GPx1 (−/−) mouse fibroblasts showed a decrease in protein kinase B (Akt) phosphorylation at Ser-473 during stimulation with hydrogen peroxide, while GPx1-under-expressed MCF-7 cells did not affect the expression and phosphorylation of p38 MAPK [52]. Homocysteine, a risk factor for cardiovascular disease, interferes with the translational reading of SECIS in the expression of GPx1 [53] and, therefore, inhibits the expression of GPx1 promoting the increase of oxygen species reagents that inactivate nitric oxide and cause endothelial dysfunction.

6. Physiopathological functions of GPx and associated diseases

The increase of ROS has been associated with the appearance and progression of aging and related diseases including arthritis, diabetes, dementia, cancer, atherosclerosis, and vascular diseases, which are inflammatory disorders, a consequence of oxidative stress [54, 55]. In reproductive medicine, free radicals cause fragmentation of spermatozoa in humans or the occurrence of ovarian failure in women, thus reducing the mobility of spermatozoa and their ability to fertilize especially in the elderly [56]. Deficiency of GPx results in direct tissue damage and activation of age-related NF-κB inflammatory pathways [57]. The application of over-expressing or knock-out and transgenic GPx1 mouse models overwhelming in vivo evidence for the protective role of GPx1 against oxidative injury and death induced by ROS and RNS. Also, the impairment of GPx1 expression is associated with the etiology of a number of chronic diseases, including cancer, cardiovascular diseases, autoimmune diseases, and diabetes [19, 27].

6.1 Diabetes

Type I diabetes, Type II DM, and gestational diabetes are characterized by hyperglycemia, dyslipidemia, and insulin resistance, which increase oxidative stress and activate the protein kinase C (PKC) as well as the receptor for advanced product glycation (AGE) and low levels of antioxidants and GPx in diabetic patients. GSH is a constituent of blood plasma. It has been found that in normal subjects, GSH plays an important role in controlling the production of free radicals, but in the case of diabetes mellitus, there is abnormal generation and elimination of plasma GSH [58]. In fact, diabetes induces an alteration in the activity of glutathione peroxidase and reductase to maintain a normal GSH level in order to avoid the increase of nitric oxide and the risk of thrombosis. However, the free radicals may play a pathogenic role in the pathophysiology of the response of glucose in β-cells and in the genesis of chronic complications. Mitochondria, the main source of ROS production, contribute to the complications of diabetes [19]. Insulin resistance, associated with mitochondrial dysfunction and increased production of ROS, alters the cardiovascular, renal, and neural functions of insulin and is a risk factor for microvascular disease. It should also be noted that ROS generates a metabolic syndrome due to changes in energy metabolism, activation of RNS, xanthine oxidase, increased expression of inflammatory mediators, and low levels of GPx and other antioxidant enzymes. Their increase induces endothelial lesions and the oxidation of LDL and redox-sensitive genes, reaching the monocyte-1 chemo attracting protein and the vascular cell adhesion molecules, molecular mechanisms that are involved in the development of the atherosclerosis. Under these conditions, it is likely that ROS and RNS contribute to the destruction of pancreatic β-cells during type diabetes. Increased levels of saturated fatty acids (FFA) and glucose in the blood are considered to be major mediators of signals that bind β-cells to apoptosis and death for T2 DM. The ER-resident GPx7 or GPx8 isoforms protect β-cells of insulin-secreting INS against lipotoxicity by enhancing the antioxidant capacity of ER without compromising insulin production and oxidative protein folding mechanisms [19, 59]. Presumably, oxidative stress is involved in the pathogenesis and complications associated with all three types of DM, and GPx1 plays a critical role in the regulation of oxidative stress [60].

6.2 GPx and obesity

Obesity promotes the storage of triglycerides in adipose tissue. Firstly, adipose tissue produces interleukin-6 (IL-6) that stimulates the absorption of dopamine creating a feeling of satiety, which has a direct effect on weight control. Tumor necrosis factor-α (NF-α) activates the NF-κB, which promotes the adhesion to the surface of endothelial and vascular smooth muscle cells of molecules causing an inflammatory state of adipose tissue, dysfunction of the endothelium, and finally, atherogenesis [61]. With the production of adipokines and decreased activity of GPx and antioxidant capacity, the endothelium becomes deficient in nitric oxide (NO), a vasodilator, and thus promotes atherosclerotic diseases. Secondly, the low level of serum GPx in obese patients and the low-serum Se concentrations, associated with the onset of signs of metabolic syndrome, may be related to the presence of a predisposing state to atherosclerosis manifested by increased consumption of antioxidants by radical interaction [19, 62].

6.3 Cardiovascular diseases

Cardiovascular diseases (CVDs) are also characterized by insulin resistance, a pro-oxidative and pro-inflammatory state, as well as a dysregulation of the expression of various factors responsible for the homeostasis of redox and inflammatory environment [62]. This is a result of oxidative stress, because plasma total GSH content is low in patients with cardiovascular disease.

High levels of homocysteine, with the slowing of GPx1 blood vessel activity, promote a higher concentration of intracellular peroxides that enhances oxidative stress and causes damage to endothelial cells in the pathogenesis of atherogenesis. Homocysteine probably interrupts UGA reading so that GPx1 expression is downregulated [53].

Studies on the evaluation of the association between GPx1 and atherosclerosis variants in Japanese patients with type 2 diabetes, with four polymorphisms, reported that functional variants of the GPx1 gene are associated with increased mean intima-media (IMT) thickness of carotid arteries and cardiovascular risk and peripheral vascular disease in type 2 diabetics [63]. These results suggest that GPx1 protects against atherogenesis in blood vessels and virus-induced myocarditis by reducing ROS levels. Disturbances in GSH metabolism may explain an increase in blood pressure related to age [64]. Selenoprotein polymorphisms are a risk factor for the development of systolic heart failure (HF) and peripheral atherosclerosis but prevent the development of abdominal aortic aneurysm (AAA). Excess weight can reduce the effectiveness of antioxidant stores in AAAs.

6.4 Neurodegenerative diseases

Neurodegenerative disorders are characterized by ROS activation of microglia that act as macrophages in the brain. The latter generate, in these glial cells, reactive nitrogenous species including inducible nitric oxide synthase (iNOS) and NADPH oxydase (NOX2); the activation of which can lead to a respiratory explosion of superoxide flooding the mitochondria and contributing more to neurodegeneration [65, 66]. GPx1 has a 10-fold higher activity in glial cells than in any other region of the brain. The in vivo administration of GPx1 to dopaminergic neurons decreases the toxicity of 6-hydroxydopamine in Parkinson's disease. By employing a lentivirus-based system to provide GPx1 to neuroblastoma cells in vitro, Ridet et al. witnessed a doubled GPx1 expression that protected the neuroblastoma cells against 6-hydroxy-dopamine-induced neurotoxicity. Other studies have shown that selenocysteine in GPx significantly delays human amyloid-Î²-induced paralysis, which is positively correlated with the incidence of Alzheimer's disease and recovers β-amyloid-induced toxicity and reduces the cellular level of EOS, by positively affecting life span and age-related pathophysiological alterations [67, 68]. Kainic acid is a neurodegenerative drug that induces PN formation in the brain. GPx1 knock-out mice are more resistant to kainic acid-induced mortality and seizures than wild-type mice. It is likely that the roles of GPx1 in neurodegenerative diseases are specific as appropriate [66, 69].

6.5 Autoimmune diseases

Generally, people infected with HIV have low levels of Se and GPx1 activity. Analysis of 75S labeling of Jurkat human T cells revealed four 75S proteins including GPx1, GPx4, TR1, and Sep15. Taking into account the function of these selenoproteins, we can think that Se influences the pathogenesis of acquired immune deficiency syndrome (AIDS) via redox regulation. The possible mechanism is that GPx1 protects HIV-infected individuals from the loss of helper T cells by preventing oxidative-induced apoptosis. HIV replication depends on the activation of NF-κB [70].

6.6 Cancers

GPx1 has an impact on signal transmission related to cell death, protein kinase phosphorylation, and activation of NF-κB via an oxidant; the anomaly of the expression of its activity would be at the base of several diseases notably cancer and chronic diseases [71]. It is known that the single nucleotide polymorphism (SNP) that alters the sequences of a particular amino acid of 201 amino acids, GPx1, is associated with certain diseases including lung, bladder, and breast cancers [72]. The cells of cancer patients often have defects in the regulation of proliferation, apoptosis, and senescence. DNA analysis of breast and colorectal cancers revealed that 36–42% of GPX1 genes lose heterozygosity during tumor formation [73]. Azoxymethane treatment of Sec-tRNA$^{[Ser]\,Sec}$ i6A transgenic mice with reduced expression of GPx1 resulted in aberrant crypts in their colon compared to wild-type mice. It appears that the SNP of the GPx1 Pro198Leu would be influenced by modulation of ROS levels and the regulation of carcinogenesis. Additional identification of GPx tagSNPs and systematic evaluation of their associations with cancer will help to expand the ability to diagnose and treat GPx1-related cancers. The GPx1 allelic loss of pathologically normal tissue adjacent to tumors would be an early event in cancer progression. Chu et al. suggested a possible protective role of GPx2 against colon cancer. GPx3 has an antioxidant protective role for proximal kidney epithelial cells in patients with kidney disease [74].

6.7 Chronic hepatopathy

Patients with chronic liver diseases have shown that disturbances of antioxidant parameters in their blood may be the cause of peroxidative damage to hepatocytes. Elevated serum carbonyl protein levels, glutathione, GPx, and glutathione reductase activities significantly decreased following increased oxidative stress in patients with pulmonary and extra pulmonary tuberculosis [75].

7. Heath benefits of traditional foods as source of precursors of glutathione and glutathione peroxidase

Hepatic GSH level is closely related to nutritional conditions, especially the cysteine content of the diet. One of the major determinants of the rate of GSH synthesis is the availability of cysteine. Cysteine is derived normally from the diet, by protein breakdown and in the liver from methionine via transulfuration [3, 11]. Se, incorporated as selenocysteine in GPx, acts in antioxidant defense and thyroid hormone formation as a protective agent against cancer, muscle diseases, coronary heart disease, and HIV [76]. In the immune system, Se stimulates the formation and activity of antibodies to helper T cells, cytotoxic T lymphocytes, and natural killer (NK) cells. The Se level drops during oxidative stress. High Se intake may be associated with a reduced risk of cancer. The recommended daily intake of Se ranges from 60 μg/day for women to 70 μg/day for men. The requirements are estimated at 100 μg/kg dry matter and 200 μg/kg for pregnant or lactating women. Generally, the dietary intake of Se ranges from 7 to 499 μg/day, with average values ranging from 40 to 134 μg/day depending on the country [70, 76, 77]. The reasons for the variability of consumption are related in particular to factors that determine the availability of selenium in the food chain, including soil pH, organic matter content, as well as the presence of ions that can form a complex with Se. Se deficiency can cause several diseases and even cause reproductive disorders in humans and animals [77]. The health benefits of Se have increased considerably since the discovery of diseases associated with polymorphisms in selenoprotein genes. Low Se status has been associated with impaired immune function with cognitive decline and increased risk of death, while Se supplementation with deficient individuals reduces the risk of prostate cancer, lung cancer, colorectal cancer, and bladder [76, 77]. Daily supplementation with Se at a supra-nutritional dose (200 μg) results in significant reductions in mortality associated with total carcinomas of lung, prostate, and colon cancers without knowing how Se reduces the risk of cancer and if the GPx1 is involved in the action [42]. Several cross-sectional studies have demonstrated the correlation between high Se status and plasma cholesterol. Indeed, Se supplementation increases the ratio of total cholesterol to significant HDL cholesterol in the plasma, suggesting a potentially beneficial effect on cardiovascular risk supplementation [77]. Prospective studies have provided some evidence of beneficial effects of Se on the risk of lung, bladder, colorectal, liver, esophagus, cardio-gastric, thyroid, and prostate cancers, but these results are sometimes disparate [77]. High Se status was associated with decreased risk of hyperglycemia and type 2 diabetes, while other studies also reported that high serum Se concentration favored increased prevalence of diabetes male participants followed over

9 years. Discrepant Se status in cancer and diabetes is thought to be associated with systemic inflammatory status and insulin resistance [77, 78]. The effects of Se on human health are multiple and complex, requiring additional research to maximize the benefits and reduce the risks of this powerful trace element. The controversial results of studies on Se supplementation based on the plasma concentrations of the individuals studied require that selenoprotein polymorphisms may be taken into account. At this stage, it should be remembered that supplemental Se intake from fortification of foods or supplements would be beneficial to people with low status, while those with adequate or high status could be negatively affected and no Se supplements would be required [77]. Tissue concentration of GSH is controlled by the availability of substrate supplied by the diet, the nutritional effects on GSH synthetic enzymes, and nutritional influences on the uptake and efflux mechanisms for GSH. Sulfur amino acid and Se contents of the diet regulate tissue GSH concentration [79]. Forms of Se in foods are essentially selenomethionine from plant sources and selenocysteine from animal sources [47].

Traditional and ethnic foods have already existed for a long time and have cultural and traditional values. Ethnic foods are defined as those edibles that are eaten and prepared by groups of people who share a common religion, language, culture, or heritage. Of course, many ethnic foods are prepared using traditional foods and vice versa [80]. Traditional foods consist of vegetables, fruits, nuts, seeds, yams, mushrooms, herbal teas, meat, fish, insects, etc. The flora of the Democratic Republic of Congo is rich in plant resources with high nutritional values and powerful medicinal properties [81, 82]. Some of traditional foods could be considered as a source of precursors of GSH and GPx by their amino acid and Se contents. Mbemba et al. studied traditional foods from Bandundu areas of DRC and listed some edible vegetables, mushrooms, nuts, and roots that showed high nutritional value [83]. Amino acid content of certain traditional foods was interesting about the presence of considerable quantity of cysteine and methionine considered such as precursor amino acids of glutathione. *Salacia pynaerti*, *Gouania longipetala*, *Dewevrea bilabiata*, *Phytolacca dodecandra* and *Solanum macrocarpon* were the traditional vegetables identified which are rich in methionine and *Alternanthera sessilis*, *Gnetum africanum*, *Justicia* sp., *Olax subscorpioides*, and *Salacia pynaerti* and rich in cysteine. Some mushrooms were equally identified which are rich in cysteine: *Auricularia delicate*, *Cookeina sulcipes*, *Gymnopilus* sp., *Lentinus squarrosulus*, *Pilocratera engleriana*, *Pleutorus tuber-regium*, *Shizophyllum commune*, and *Oudemansiella canarii* [83]. Regarding Se, the literature indicates that the animal-derived foods tend to be a better dietary of these micronutrients. Seafoods and organ meats are the richest food sources of selenium [84]. Preliminary studies on mineral composition of Congolese traditional foods showed that edible insects are the excellent sources of Se (unpublished data). Edible insects are known to be excellent sources of mineral micronutrients [85, 86]. Other sources of Se include cereals, grains, and dairy products [84]. Selenium concentrations in plant-based foods vary widely by geographic location. For this, Brazil nuts (*Bertholletia excelsa*, H.B.K.) have the highest selenium concentration of all edible nuts and are considered one of the most selenium concentrated food sources [87, 88].

8. Conclusion

The glutathione system plays important biological roles, including the defense of cellular tissues against reactive oxygen and nitrogen species, as well as maintaining the redox status and detoxification of cells. GPx, the main enzyme in the antioxidant line, is characterized by eight isomorphs whose activities are localized in the various subcellular organelles. During aging, the GSH system in general and the GPx family in particular undergo modifications that promote the production of oxidative stress resulting in disturbances in metabolic regulation, damage of cellular constituents accompanied by cardiovascular follow-up, neurodegeneration, and cancers. In this evolution, GPx may be a parameter contributing to the identification of subclinical asymptomatic cardiometabolic disorders and their repair, since its decrease favors atherosclerotic diseases. GPx deficiency results in direct tissue damage and activation of age-related NF-κB inflammatory pathways, which is associated with aging. Data accumulates to bind alteration or abnormality of GPx1 expression toward the etiology of cancer, cardiovascular diseases, neurodegeneration, autoimmune disease, and diabetes. The involvement of the GSH system and GPxs in various diseases, especially those of the elderly, is obvious. However, at this stage, there is a need for a thorough study to better elucidate the mechanism of GPx1 in the pathogenesis and potential applications of GPx1 manipulation in the treatment of these disorders.

Acknowledgements

The authors thank Dr. Christian Isalomboto Nkanga, Jean Paul Nzundu Mbo, Priscille Mbemba Mulosi, and José Mbemba Kiniema for their contribution.

Conflict of interest

The authors declare that they have no conflict of interest.

Author details

Théophile Mbemba Fundu[1], Paulin Mutwale Kapepula[2], Justin Mboloko Esimo[3], José Remacle[4] and Nadege Kabamba Ngombe[2*]

1 Faculty of Sciences, Department of Biology, Laboratoire d'analyses et de recherches sur les aliments et la nutrition, University of Kinshasa, Kinshasa, Democratic Republic of the Congo

2 Faculty of Pharmaceutical Sciences, Centre d'Etudes des Substances Naturelles d'Origine Végétale (CESNOV), University of Kinshasa, Democratic Republic of Congo

3 Faculty of Medicine, University of Kinshasa, Democratic Republic of Congo

4 Laboratory of Cellular Biochemistry and Biology (URBC), University of Namur, Namur, Belgium

*Address all correspondence to: nadegengk@gmail.com

References

[1] Wu G, Fang Y-Z, Yang S, Lupton JR, Turner ND. Recent advances in nutritional sciences glutathione metabolism and its. The Journal of Nutrition. 2004;**134**(3):489-492

[2] Arteel GE, Sies H. The biochemistry of selenium and the glutathione system. Environmental Toxicology and Pharmacology. 2001;**10**(4):153-158

[3] Lu SC. Glutathione synthesis. Biochimica et Biophysica Acta. 2013;**1830**(5):3143-3153

[4] Halliwell B, Gutteridge JMC. Free Radicals in Biology and Medicine. 5th ed. Oxford University Press; 2015. 961 p

[5] Farhat Z, Browne RW, Bonner MR, Tian L, Deng F, Swanson M, et al. How do glutathione antioxidant enzymes and total antioxidant status respond to air pollution exposure? Environement International. 2018;**112**:287-293

[6] Mbemba T, Houbion A, Remacle J. Alteration of enzymes in ageing human fibroblasts in culture IV. Effect of glutathione on the alteration of glucose-6.phosphate dehydrogenase. Mechanisms of Ageing and Development. 1985;**32**:249-266

[7] Mischley LK, Lau RC, Shankland EG, Wilbur TK, Padowski JM. Phase IIb study of intranasal glutathione in Parkinson's disease. Journal of Parkinson's Disease. 2017;**7**(2):289-299

[8] Montero D, Tachibana C, Rahr Winther J, Appenzeller C. Intracellular glutathione pools are heterogeneously concentrated. Redox Biology. 2013;**1**(1):508-513

[9] Aquilano K, Baldelli S, Ciriolo MR. Glutathione: New roles in redox signalling for an old antioxidant. Frontiers in Pharmacology. 2014;**5**:1-12

[10] Limon-Pacheco JH, Gonsebatt ME. The glutathione system and its regulation by neurohormone melatonin in the central nervous system. Central Nervous System Agents in Medicinal Chemistry. 2012;**10**(4):287-297

[11] Lu SC. Regulation of glutathione synthesis. Current Topics in Cellular Regulation. 2000;**36**:95-116

[12] Butterfield D, Castegna A, Lauderback C, Drake J. Oxidative stress and neurodegenerative diseases: A review of upstream and downstream antioxidant therapeutic options. Current Neuropharmacology. 2009;**7**(1):65-74

[13] Bajic VP, Van Neste C, Obradovic M, Zafirovic S, Radak D, Bajic VB, et al. Glutathione "redox homeostasis" and its relation to cardiovascular disease. Oxidative Medicine and Cellular Longevity. 2019;**2019**:1-14

[14] Lavoie M-E, Lhoret RR, Ziai S, Lavoie J-C. Blood glutathione peroxidase activity in relation with the risk of cardiovascular diseases in obese women. Journal of Diabetes and Metabolism. 2011;**2**(05):2-6

[15] Meredith MJ, Redd DJ. Status of the mitochondrial pool of glutathione in the isolated hepatocyte. The Journal of Biological Chemistry. 1982;**257**(7):3747-3753

[16] Kumar C, Igbaria A, D'Autreaux B, Planson AG, Junot C, Godat E, et al. Glutathione revisited: A vital function in iron metabolism and ancillary role in thiol-redox control. The EMBO Journal. 2011;**30**(10):2044-2056

[17] Morales A, Colell A, García-ruiz C, Kaplowitz N, Fernández-checa JC. Mitochondrial glutathione: Features,

regulation and role in disease. Biochimica et Biophysica Acta. 2012;**4**:1-12

[18] Mbemba F, Houbion A, Raes M, Remacle J. Subcellular localization and modification with ageing of glutathione, glutathione peroxidase and glutathione reductase activities in human fibroblasts. Biochimica et Biophysica Acta. 1985;**838**(2):211-220

[19] Brigelius-flohé R, Maiorino M. Glutathione peroxidases. Biochimica et Biophysica Acta. 2013;**1830**:3289-3303

[20] Deponte M. Glutathione catalysis and the reaction mechanisms of glutathione-dependent enzymes. Biochimica et Biophysica Acta. 2013;**1830**:3217-3266

[21] Margis R, Dunand C, Teixeira FK, Margis-PinheiroM.Glutathioneperoxidase family—An evolutionary overview. The FEBS Journal. 2008;**275**(15):3959-3970

[22] Nguyen VD, Saaranen MJ, Karala AR, Lappi AK, Wang L, Raykhel IB, et al. Two endoplasmic reticulum PDI peroxidases increase the efficiency of the use of peroxide during disulfide bond formation. Journal of Molecular Biology. 2011;**406**(3):503-515

[23] Pastori D, Pignatelli P, Farcomeni A, Menichelli D, Nocella C, Carnevale R, et al. Aging-related decline of glutathione peroxidase 3 and risk of cardiovascular events in patients with atrial fibrillation. Journal of the American Heart Association. 2016;**5**:1-9

[24] Ekoue DN, He C, Diamond AM, Bonini MG. Manganese superoxide dismutase and glutathione peroxidase-1 contribute to the rise and fall of mitochondrial reactive oxygen species which drive oncogenesis. Biochimica et Biophysica Acta. 2017;**1858**(8):628-632

[25] Esworthy RS, Aranda R, Martín MG, Doroshow JH, Binder SW, Chu F-F. Mice with combined disruption of Gpx1 and Gpx2 genes have colitis. American Journal of Physiology. Gastrointestinal and Liver Physiology. 2017;**281**(3):G848-G855

[26] Brigelius-Flohe R. Tissue-specific functions of individual glutathione peroxidases. Free Radical Biology and Medicine. 1999;**27**(9):951-965

[27] Maher P. The effects of stress and aging on glutathione metabolism. Ageing Research Reviews. 2005;**4**(2):288-314

[28] Low S. Knowing when not to stop: Selenocysteine incorporation in eukaryotes. Trends in Biochemical Sciences. 2002;**21**(6):203-208

[29] Lammi MJ, Qu C. Selenium-related transcriptional regulation of gene expression. International Journal of Molecular Sciences. 2018;**19**(9)

[30] Chada S. Post-transcriptional regulation of glutathione peroxidase gene expression by selenium in the HL-60 human myeloid cell line. Blood. 2015;**74**(3):930-939

[31] Tan M, Li S, Swaroop M, Guan K, Oberley LW, Sun Y. Transcriptional activation of the human glutathione peroxidase promoter by p53. The Journal of Biological Chemistry. 1999;**274**(17):12061-12066

[32] Puzio-Kuter AM. The role of p53 in metabolic regulation. Genes and Cancer. 2011;**2**(4):385-391

[33] Lee HT, Emala CW. Adenosine attenuates oxidant injury in human proximal tubular cells via A 1 and A 2a adenosine receptors. American Journal of Physiology. Renal Physiology. 2002;**282**(5):F844-F852

[34] Cao C, Ren X, Kharbanda S, Koleske A, Prasad KVS, Kufe D. The

ARG tyrosine kinase interacts with Siva-1 in the apoptotic response to oxidative stress. The Journal of Biological Chemistry. 2001;**276**(15):11465-11468

[35] Jozwiak Z, Jasnowska B. Changes in oxygen-metabolizing enzymes and lipid peroxidation in human erythrocytes as a function of age of donor. Mechanisms of Ageing and Development. 1985;**32**(1):77-83

[36] Habdous M, Herbeth B, Vincent-Viry M, Lamont JV, Fitzgerald PS, Visvikis S, et al. Serum total antioxidant status, erythrocyte superoxide dismutase and whole-blood glutathione peroxidase activities in the stanislas cohort: Influencing factors and reference intervals. Clinical Chemistry and Laboratory Medicine. 2003;**41**(2):209-215

[37] Walston J, Xue Q, Semba RD, Ferrucci L, Cappola AR, Ricks M, et al. Serum antioxidants, inflammation, and total mortality in older women. American Journal of Epidemiology. 2006;**163**(1):18-26

[38] Espinoza SE, Guo H, Fedarko N, DeZern A, Fried LP, Xue QL, et al. Glutathione peroxidase enzyme activity in aging. Journal of Gerontology. 2008;**63**(5):505-509

[39] Bermano G, Nicol F, Dyer JA, Sunde RA, Beckett GJ, Arthur JR, et al. Tissue-specific regulation of selenoenzyme gene expression during selenium deficiency in rats. The Biochemical Journal. 1995;**311**(2):425-430

[40] Peters U, Takata Y. Selenium and the prevention of prostate and colorectal cancer. Molecular Nutrition and Food Research. 2008;**52**(11):1261-1272

[41] Jiang C, Wang Z, Lü J, Ganther H. Distinct effects of methylseleninic acid versus selenite on apoptosis, cell cycle, and protein kinase pathways in DU145 human prostate cancer cells. Molecular Cancer Therapeutics. 2002;**1**(12):1059-1066

[42] Novoselov SV, Calvisi DF, Labunskyy VM, Factor V, Carlson BA, Fomenko DE, et al. Selenoprotein deficiency and high levels of selenium compounds can effectively inhibit hepatocarcinogenesis in transgenic mice. Oncogene. 2005;**24**(54):8003-8011

[43] Low SC, Grundner E, Harney J, Berry M. SECIS-SBP2 interactions dictate selenocysteine incorporation efficiency and selenoprotein hierarchy. The EMBO Journal. 2000;**19**:6882-6890

[44] Gonzalez-Flores JN, Gupta N, DeMong LW, Copeland PR. The selenocysteine-specific elongation factor contains a novel and multi-functional domain. The Journal of Biological Chemistry. 2012;**287**(46):38936-38945

[45] Carlson BA, Xu XM, Gladyshev VN, Hatfield DL. Selective rescue of selenoprotein expression in mice lacking a highly specialized methyl group in selenocysteine tRNA. The Journal of Biological Chemistry. 2005;**280**(7):5542-5548

[46] Gonzalez-Flores JN, Shetty SP, Dubey A, Copeland PR. The molecular biology of selenocysteine. Biomolecular Concepts. 2013;**4**(4):349-365

[47] Labunskyy VM, Hatfield DL, Gladyshev VN. Selenoproteins: Molecular pathways and physiological roles. Physiological Reviews. 2014;**94**(3):739-777

[48] Cao C, Leng Y, Huang W, Liu X, Kufe D. Glutathione peroxidase 1 is regulated by the c-Abl and Arg tyrosine kinases. The Journal of Biological Chemistry. 2003;**278**(41):39609-39614

[49] Chen F, Castranova V. Nuclear factor-κB, an unappreciated tumor suppressor. Cancer Research. 2007;**67**(23):11093-11098

[50] Chen F, Castranova V, Shi X. New insights into the role of nuclear factor-κB in cell growth regulation. The American Journal of Pathology. 2001;**159**(2):387-397

[51] Hattori H, Imai H, Hanamoto A, Furuhama K, Nakagawa Y. Up-regulation of phospholipid hydroperoxide glutathione peroxidase in rat casein-induced polymorphonuclear neutrophils. Biochemistry Journal. 2005;**389**(3):920

[52] Ono K, Han J. The p38 signal transduction pathway activation and function. Cellular Signaling. 2000;**12**:1-13

[53] Handy DE, Zhang Y, Loscalzo J. Homocysteine down-regulates cellular glutathione peroxidase (GPx1) by decreasing translation. The Journal of Biological Chemistry. 2005;**280**(16):15518-15525

[54] Weidinger A, Kozlov AV. Biological activities of reactive oxygen and nitrogen species: Oxidative stress versus signal transduction. Biomolecules. 2015;**5**(2):472-484

[55] López-Otín C, Blasco MA, Partridge L, Serrano M, Kroemer G. The hallmarks of aging. Cell. 2013;**153**(6):1194

[56] Opuwari CS, Henkel RR. An update on oxidative damage to spermatozoa and oocytes. BioMed Research International. 2016;**2016**:1-11

[57] Erden-Inal M, Sunal EKG. Age-related changes in the glutathione redox system. Cell Biochemistry and Function. 2002;**20**:61-66

[58] Wautier M-P, Chappey O, Corda S, Stern DM, Schmidt AM, Wautier J-L. Activation of NADPH oxidase by AGE links oxidant stress to altered gene expression via RAGE. American Journal of Physiology. Endocrinology and Metabolism. 2017;**280**(5):E685-E694

[59] Paneni F, Beckman JA, Creager MA, Cosentino F. Diabetes and vascular disease: Pathophysiology, clinical consequences, and medical therapy: Part I. European Heart Journal. 2013;**34**(31):2436-2446

[60] Tatsch E, Carvalho JAMD, Hausen BS, Bollick YS, Torbitz VD, Duarte T, et al. Oxidative DNA damage is associated with inflammatory response, insulin resistance and microvascular complications in type 2 diabetes. Mutation Research, Fundamental and Molecular Mechanisms of Mutagenesis. 2015;**782**:17-22

[61] Lastra G, Manrique CM, Hayden MR. The role of beta-cell dysfunction in the cardiometabolic syndrome. Journal of the Cardiometabolic Syndrome. 2006;**1**(1):41-46

[62] Roberts CK, Sindhu KK. Oxidative stress and metabolic syndrome. Life Sciences. 2009;**84**(21-22):705-712

[63] Zalesin KC, Franklin BA, Miller W, Peterson ED, McCullough PA. Impact of obesity on cardiovascular disease. The Medical Clinics of North America. 2011;**95**(5):919-937

[64] Robaczewska J, Kedziora-Kornatowska K, Zary-Sikorska E, Pawluk H, Pawliszak W, et al. Role of glutathione metabolism and glutathione-related antioxidant defense systems in hypertension. Journal of Physiology and Pharmacology. 2016;**67**(3):331-337

[65] Solleiro-Villavicencio H, Rivas-Arancibia S. Effect of chronic oxidative stress on neuroinflammatory response mediated by CD4+T cells in neurodegenerative diseases. Frontiers in Cellular Neuroscience. 2018;**12**(April):1-13

[66] Magalingam KB, Radhakrishnan A, Ping NS, Haleagrahara N. Current concepts of neurodegenerative mechanisms in Alzheimer's disease. BioMed Research International. 2018;**2018**:1-12

[67] Kim SH, Kim BK, Park SK. Selenocysteine mimics the effect of dietary restriction on lifespan via SKN-1 and retards age-associated pathophysiological changes in *Caenorhabditis elegans*. Molecular Medicine Reports. 2018;**18**(6):5389-5398

[68] Ridet JL, Bensadoun JC, Déglon N, Aebischer P, Zurn AD. Lentivirus-mediated expression of glutathione peroxidase: Neuroprotection in murine models of Parkinson's disease. Neurobiology of Disease. 2006;**21**(1):29-34

[69] Yon JM, Kim YB, Park D. The ethanol fraction of white rose petal extract abrogates excitotoxicity-induced neuronal damage in vivo and in vitro through inhibition of oxidative stress and proinflammation. Nutrients. 2018;**10**(10)

[70] Avery JC, Hoffmann PR. Selenium, selenoproteins, and immunity. Nutrients. 2018;**10**(9):1-20

[71] Metere A, Frezzotti F, Graves C, Vergine M, De Luca A, Pietraforte D, et al. A possible role for selenoprotein glutathione peroxidase (GPx1) and thioredoxin reductases (TrxR1) in thyroid cancer: Our experience in thyroid surgery. Cancer Cell International. 2018;**18**(1):1-8

[72] Ichimura Y, Habuchi T, Tsuchiya N, Wang L, Oyama C, Sato K, et al. Increased risk of bladder cancer associated with a glutathione peroxidase 1 codon 198 variant. The Journal of Urology. 2004;**172**(2):728-732

[73] Hu Y, Benya RV, Carroll RE, Diamond AM. Allelic loss of the gene for the GPX1 selenium-containing protein is a common event in cancer. The Journal of Nutrition. 2005;**135**(12):3021S-3024S

[74] Ulrich Ulrike H, Elizabeth MP, Xiao L, Curtin K, Duggan D, Hsu L, et al. Glutathione peroxidase tagsSNPs: Associations with rectal cancer but not with colon cancer. Genes, Chromosomes and Cancer. 2012;**47**:238-246

[75] Davis R, Emerling BM, Platanias LC, Black E, Nebreda AR, Davis RJ, et al. Mitochondrial reactive oxygen species activation of p38 mitogen-activated protein kinase is required for hypoxia signaling. Molecular and Cellular Biology. 2015;**25**(June):4853-4862

[76] Tan HW, Mo HY, Lau AT, Xu YM. Selenium species: Current status and potentials in cancer prevention and therapy. International Journal of Molecular Sciences. 2019;**20**(1)

[77] Rayman MP. Selenium and human health. Lancet. 2012;**379**(9822):1256-1268

[78] Steinbrenner H, Bodo S, Pinto SH. High selenium intake and increased diabetes risk: Experimental evidence for interplay between selenium and carbohydrate metabolism. Journal of Clinical Biochemistry and Nutrition. 2011;**49**(3):174-181

[79] Bray TM, Taylor CG. Tissue glutathione, nutrition, and oxidative stress. Canadian Journal of Physiology and Pharmacology. 2011;**71**(9):746-751

[80] Prakash V, Martin-Belloso O, Astley SB, Braun S, Mcmahon H, Lelieveld H. Regulating Safety of Traditional and Ethnic Foods. 2016. 538 p

[81] Kapepula PM, Tshala-Katumbay D, Mumba ND, Frédérich M, Mbemba FT, Kabamba NN. Traditional foods as putative sources of antioxidants with health benefits in Konzo. In: Antioxidants in Foods and Its Applications. 2018. p. 182

[82] Bahati LM, Kapepula PM, Kabamba N, Moni B, Kafuti G, Mungitshi M, et al. Microscopic features, chromatographic fingerprints and antioxidant property of some unconventional green leafy vegetables consumed in Bandundu, DR Congo. Pharmacognosy Communications. 2017;7(4):158-163

[83] Mbemba FT, Remacle J, Paulus J, Anita A, Timi-Timi A, Kwilu N. Aliments et denrées alimentaires traditionnels du Bandundu en R. D. Congo. L'Harmatan: Répertoire et composition en nutriments; 2013. p. 317

[84] Stoffaneller R, Morse NL. A review of dietary selenium intake and selenium status in Europe and the Middle East. Nutrients. 2015;7(3):1494-1537

[85] Kapepula PM, Ngumbi PB, Kawayidiko MK, Mpanzu AD, Franck T, Mouithys-mickalad A, et al. Anti-inflammatory and antioxidant activities of Rungia congoensis, a traditional vegetable consumed by Yombe people from Kongo Central area (DR. Congo). Natural Product Research. 2018;**6419**:1-5

[86] Malaisse F. Se nourrir en fôret claire africaine: Approche écologique et nutritionnelle. Les presses agronomiques de Gembloux; 1997. p. 384

[87] Cardoso BR, Duarte GBS, Reis BZ, Cozzolino SMF. Brazil nuts: Nutritional composition, health benefits and safety aspects. Foodservice Research International. 2017;**100**:9-10

[88] Cominetti C, de Bortoli MC, Garrido AB, Cozzolino SMF. Brazilian nut consumption improves selenium status and glutathione peroxidase activity and reduces atherogenic risk in obese women. Nutrition Research. 2012:403-407

Chapter 3

Glutathione Peroxidase in Health and Diseases

Eren Sarıkaya and Selami Doğan

Abstract

The aim of this study is to give information to readers about the importance of glutathione peroxidase. The physiopathology of most diseases is not fully elucidated currently; however, in many epidemiological studies, there are limited studies indicating the relationship between low levels of glutathione peroxidase status and the rise of cancer risk in many types of cancer. Anytime, situations in case of the distortion due to imbalance between enzymatic and nonenzymatic antioxidants and oxidants which lost one of paired electrons in the atomic level mean reactive oxygen species (ROS) withal reactive nitrogen species (RNS) in favor of oxidants that are related to oxidative stress. The possible mechanisms of glutathione peroxidase have been reviewed using the major findings of more than 1000 papers related to the ROS, glutathione peroxidase, and oxidative stress. Oxidative stress plays an important role in the occurrence and development of most diseases in both animal and human studies. Moreover, antioxidants have protective effects against nearly 50 disease pathogenesis. Oxidative stress, which occurs as an outcome of lipid peroxidation, concurrently may have a key importance in the phase of carcinogenesis occurring with a multistage course devoted to environmental toxicity and in cancer pathogenesis.

Keywords: glutathione peroxidase, oxidative stress, antioxidant, free radical

1. Introduction

1.1 Reactive oxygen species

Reactive oxygen species (ROS), superoxide radicals (O_2-), hydrogen peroxide (H_2O_2), hydroxyl radical (OH-), and singlet oxygen are formed in normal oxygen metabolism. Free radicals can initiate free radical sequence reactions that shape various free radicals [1]. Reactive oxygen species have been associated with many disease categories, including cancer. In addition, ROSs have been reported to increase tumor cell migration and increase the risk of metastasis and metastasis. It is known that the harmful effects of ROS are controlled by various cellular defense systems consisting of enzymatic components (catalase, glutathione peroxidase and superoxide dismutase, etc.). Epidemiological literature studies have found a relationship between low levels of antioxidants and an increased risk of cancer [2].

Although many definitions are made for free radicals, in the general definition, free radical is a chemical product that is in molecular or atomic orbit and is generally highly reactive and contains unpaired electrons. The electrons in atoms move in spaces called orbit. Each orbit has up to two electrons moving in opposite directions. Free radicals can be positively charged, negatively charged, or neutral. In biological systems, they occur mostly by electron transfer. Although free radical reactions are necessary for the defense mechanism of cells such as neutrophils and macrophages from immune system cells, overproduction of free radicals results in tissue damage and cell death [3].

Electrons are orbital in atoms and are present in pairs in the spatial region. Bonds are formed as a result of the interaction between atoms and molecular structure is formed due to these bonds. Free oxygen radicals, atomic or molecular structures in the single electron, are the name given to uncommon parts. These molecules that easily exchange electrons with other molecules are also called free oxygen radicals (FOR) or reactive oxygen radicals (ROR) [4].

As a result of oxidation which is a natural process during the process of metabolism, the formation of free radicals, which cause various damages in the organism and initiator of some vital chronic diseases such as cancer and heart diseases, increased the interest in antioxidant compounds that combat them, and studies on this subject mostly focused on the determination of antioxidant activities. This interest is directed mainly to reactive oxygen species (ROS) damage in the aging process and in the etiopathogenesis of many diseases. Aging and chronic diseases in humans occur as a result of some complex biological processes. To understand these complex processes, various hypotheses have been proposed and tested experimentally. The theories about aging have been explained in recent years by some advances in molecular genetics and experimental techniques. Increasing damage caused by ROS in the cell mainly involves telomere erosion, genome instability, DNA mutations, and changes in gene profiles in senescent cells [5, 6].

The formation of free radicals occurs during the use of oxygen in the organism. Unmapped electron-containing atoms or molecules initiate a sequence of reactions in which cells are damaged. The formation of free radicals in the body begins and increases with catabolic reactions as well as factors such as fatty diets, unhealthy nutrition, smoking, drug treatments, alcohol consumption, radiation, pesticides, and environmental pollution. Free radicals weaken the immune system, leading to various diseases and premature aging. In this respect, antioxidants are important in cell-protective treatment and protection from degenerative diseases. Research has shown that antioxidants neutralize free radicals and prevent damage to cells. In the context of antioxidant compounds, microalgae species have an important place besides terrestrial foods. When some microalgae species are grown under various stress conditions (nitrogen deficiency, high light intensity, high salinity, etc.), it is possible to accumulate in the cell pigment substances with strong antioxidant properties such as beta-carotene, astaxanthin, zeaxanthin, and lutein. Thus, in the context of biotechnology, microalgalas can be manipulated with various parameters in culture conditions to create various physiological stresses on cells. Thus, it can be ensured that the cultured cells produce more of the desired product [6].

The form of lipid peroxidation with the result of molecular oxygen conversion to reactive oxygen species (ROS) with various environmental factors, particularly cigarette smoke, alcohol, UV rays, and other oxidants, leads to oxidative stress. As a result of this, a multistage carcinogenesis process is favored by ROS in the body.

1.2 Oxidative stress

Oxidative stress is an important component in binding environmental toxicity to a multistage carcinogenic process. In addition, oxidative stress is characterized by the cumulative effect of more than one activity, such as a multistage process (three stages in a single cell; onset, elevation, and progression), such as cancer development. Reactive oxygen species (ROS) are produced in response to endogenous and exogenous stimulation. ROS can affect all these stages of carcinogenesis [7]. For this reason, the term oxidative stress is used to describe the imbalance between cellular levels of oxidants and antioxidants [8].

Cell damage caused by free radicals is believed to play an important role in the progression of aging process and aging-related degenerative diseases (especially atherosclerosis, cataract, diabetes, neurodegenerative diseases, immunosystem disorders, and cancer formation). Oxidative stress has been associated with almost 50 disease pathogenesis [9].

Catalase (CAT), peroxidase (POD), glutathione reductase (GR), and superoxide dismutase (SOD) are enzymes that have antioxidant effects in biological and biochemical systems. The antioxidant defense system protects the cell against oxidative damage of free radicals or other reactive molecules. Therefore, antioxidant enzymes such as CAT, POD, GSSG-Rx, and SOD are of great importance in this defense system. The harmful effects of free radicals are controlled by antioxidant defense systems in cells [10].

1.3 Glutathione peroxidase

Antioxidants are substances that prevent, reduce, or delay the oxidation of materials that may be exposed to oxidation such as proteins, lipids, carbohydrates, and DNA in living cells, and this is called antioxidant defense. Antioxidants are substances that prevent or delay the damage of free oxygen radicals on target tissues. Antioxidants are classified into two categories, enzymatic and nonenzymatic. Enzymatic antioxidants are superoxide dismutase (SOD), catalase, and glutathione peroxidase (GPx), and nonenzymatic antioxidants are vitamin E, vitamin C, vitamin A, selenium (Se), transferrin, and lactoferrin. Antioxidants are often intracellular and sometimes extracellular [4]. Reduction in circulating antioxidant may be due to sequestration by tumor cells as well as sweeping lipid peroxides by tumor cells [2].

The negative aspect of ROS production is that ROS constitutes various types of cancer that can be resistant to exogenous growth in itself [11, 12]. For example, HL-60, multidrug-resistant, is resistant to growth of ROS due to the endogenous height of antioxidants which are detoxifying and which are ROS scavenger such as leukemia and CAT [11, 12]. Several oncogene-induced cancer cells enhance the antioxidant activity by activating nuclear factor erythroid 2-related factor 2 (NRF2) and maintaining the effect [12–16]. ROS levels allow the activation of pro-tumorigenic signaling pathways without induced cell death [12, 17]. Also, in the event of an increase in GSH levels, which play an active role in protecting cells from cell death, also appears to play an active role in protection from ROS-inducing therapy [12, 17]. The antioxidant defense system is a large network of molecules which eliminate free radicals and production of ROS [18, 19]. Endogenous antioxidant defense systems are available to compensate the ROS welded damage [7, 20]. These systems function their form by maintaining intracellular ROS activity and redox balance with chelating [7, 20].

Glutathione peroxidase catalyzed by the reaction catalyzed by the reduced form of glutathione (GSH) by reacting with hydrogen peroxide or lipid peroxides while playing a role in the detoxification of these molecules by creating a glutathione bridge with another glutathione molecule (GSSG) form [21]. H_2O_2 is detoxified by catalase and glutathione peroxidase [22].

The glutathione redox cycle plays a key role in the reduction of intracellular hydroperoxides. GPx belongs to the class of selenocysteine compound because it binds four atoms of selenium and provides the catalytic activity of glutathione peroxidase. It needs glutathione as a co-substrate [23].

Glutathione is a tripeptide which is composed of cysteine, glutamic acid, and glycine. GSH has two structural characteristic: γ-glutamyl linkage and sulfhydryl (-SH) group. GSH is known for its multiple physiological functions as an antioxidant against ROS and free radicals in detoxification of xenobiotic compounds [24–28]. When the cell fails to protect GSH content no longer, certain cell death may be followed [28, 29]. GSH, which is the most important antioxidant molecule of intracellular environment, has many physiological functions such as detoxification of xenobiotics, transport of amino acids, keeping sulfhydryl groups in proteins in the reduced state, and acting as a coenzyme in some enzymatic reactions other than involving antioxidant defense system [21, 30–32]. Glutathione, in reduced form (GSH), turns itself into oxidized glutathione (GSSG) form with creating disulfide bridge with another glutathione molecule while playing role in detoxification of these molecules by reacting with hydrogen peroxides or lipid peroxides, with the reaction catalyzed by the GPx enzyme. In order for the maintenance of free radical detoxification in cell, GSSG needs to be converted back to the reduced form. GSSG is converted to reduced glutathione form with GR enzyme by a reaction in which NADPH is used [21, 30, 33].

Glutathione peroxidase (GSH-Px; E.C. 1.11.1.9) is a cytosolic enzyme responsible for the reduction of hydroperoxides. In erythrocytes, GSH-Px is the most effective antioxidant against oxidant stress and has some important functions in phagocytic cells [34]:

$$H_2O_2 + 2GSH \xrightarrow{GP_x} 2H_2O + GSSG \tag{1}$$

$$ROOH + 2GSH \xrightarrow{GP_x} ROH + GSSG + H_2O \tag{2}$$

GPx is an enzyme responsible for the removal of hydroperoxides formed in cells. Since its subunits contain a Se atom, it is thought to be a selenoenzyme that protects cells against various damages. The presence of this enzyme was first found by Mills in 1957 in mammalian erythrocytes. It is the most effective enzyme in endothelial cells and especially in the lung. About 60–75% of enzyme activity is found in the eukaryotic cell cytoplasm and 25–40% is found in mitochondria. The most common enzyme activity is clear in erythrocytes and liver. GPx is the most important enzyme that protects lipids from peroxidation at intracellular distance. For this reason, this enzyme, especially in the cytosolic compartment of the cell, maintains the structure and function of the cell. The phospholipid hydroperoxide glutathione peroxidase (PLGSH-Px), which reduces membrane phospholipid hydroperoxides to alcohol, also contains the Se atom and is monomeric. It is also a cytosolic enzyme. PLGSH-Px provides protection against peroxidation of the membrane in cases where vitamin E, a membrane-bound antioxidant, is insufficient [35–37].

Glutathione peroxidase catalyzes detoxification of H_2O_2 and lipid peroxides by reduced glutathione. Thus, it protects membrane lipids and hemoglobin from oxidation of peroxides. GSH-Px is also involved in the detoxification of xenobiotics. It is the antioxidant enzyme system that provides the most vital defense against the peroxidative damage of biological membranes in mammalian cells. From these enzymes, glutathione peroxidase, catalase, and superoxide dismutase together form a common system aimed at protecting the cell from peroxidant molecules [38].

In a study, when the serum GPx ($p < 0.001$) levels were compared, the coronary artery ectasia (CAE) patient group was significantly lower compared to the control group [39]. In a literature study, there was no significant difference

($p > 0.05$) between GPx activities in both groups of patients with malignant tumors in the head and neck region [40]. An increase in the level of free oxygen radicals may cause mutagenicity, cytotoxicity, and changes in gene expression, leading to the development of malignant tumors, and this mutagenicity may contribute to the transformation of benign development into malignant [41]. In conclusion, high GPx activity in the blood can be considered as a factor reducing cancer risk. Comparison of oxidative parameters in Parkinson's disease groups and control groups showed that GSH-Px was significantly higher in the patient group [42]. Decreased levels of GPx activities were observed in progressive hypothyroidism in the postnatal period [43]. Reduced activities of GPx, one of the most important antioxidant defense systems in the body, either reported increased or decreased intracellularly in different tissues antioxidant enzymes [44, 45]. Increased GPx activity experimental colitis was reported in the parotid glands of mice; decreased enzyme activity was found in submandibular glands [46]. The results of GPx activities in plasma in both Crohn's disease (CD) groups as well as controls did not reveal any statistically significant differences [47]. Results of GPx and SOD activities measured in CD patients have been demonstrated to be diverse when analyzing plasma samples [48]. Previous studies focusing on the role of GPX1 single nucleotide polymorphisms (SNPs) included GPX1 SNP rs1050450, which often caused the C to T mutation [49]. GPX1 SNP was found to affect the risk of lung cancer and bladder cancer of r1010450 [50]. The risk of vascular calcification and atherosclerosis is also affected by the Leu allele of GPX1 SNPs [51, 52]. However, available evidence did not acknowledge that GPX1 SNP rs1050450 plays an important role in chronic kidney disease (CKD) progression or renal allograft dysfunction [53, 54]. Significant influences of GPX1 Pro197Leu SNPs on the risk of ESRD in Han Chinese population have not been detected [55]. It is suggested that individuals with low GPx activity are prone to intact antioxidant protection leading to oxidative damage to membrane fatty acid and functional proteins and consequently to neurotoxic damage [56]. Forgione and colleagues previously hypothesized that GPX1 deficiency caused an increase in vascular oxidative stress with endothelial dysfunction directly involved [57]. Inhibition of ferroptosis by GPx4 provides protective mechanisms against neurodegeneration. In addition, we suggest that selenium deficiency increases susceptibility to ferroptotic processes and other programmed cell death pathways due to a decrease in GPx activity [58]. GPX1 affects the effects of the major factors involved in both macro and micronutrient metabolism by regulating gene expression, protein function, and enzyme activities [59–61]. Some studies [62, 63] underline the importance of maintaining an appropriate expression and activity of this selenoperoxidase to control redox balance and glucose and lipid metabolism. GPX1 polymorphism with risk of diabetes and obesity in different populations [64–68]. Oxidative stress-induced intestinal injury has been reduced by the addition of SOD, glutathione peroxidase, and N-acetylcysteine, which reduces intestinal tissue tumor necrosis factor-a concentrations with anti-inflammatory and antioxidant properties [69, 70]. A significant increase in GPx activity in the inflamed mucosa was found in either active or remission stage in ulcerative colitis (UC) patients. Other studies confirmed significantly higher plasma GPx levels in the UC and CD groups than in the control group [71]. Children with inflammatory bowel diseases (IBDs) had increased GPx activity and GSH content compared with control children [72]. Selenium supplementation in patients with autoimmune thyroiditis is associated with decreased antithyroperoxidase antibody levels, improved thyroid ultrasound characteristics, and improved quality of life [73]. ROS has been incorporated into cellular signaling, which may activate mitogenic cellular pathways and proinflammatory processes leading to disruption of renal fibrosis and GPx, which support progressive impairment of renal function [74]. Total glutathione, decreased/oxidized glutathione, and ubiquinone were significantly decreased in patients with susceptibility-related diseases, while DNA fragmentation was significantly increased in patients. However, these differences were not associated with the GPx1 genetic background [75]. It has been reported a linear relationship between estrogen and GPx in erythrocytes of postmenopausal women [76]. Serviddio et al. [77] showed a positive correlation between the activity of GPx and luteinizing hormone (LH) concentrations in healthy women. They also observed a significant positive correlation between estradiol and GPx. Significant increase in GPx activity was found in abdominal obesity in postmenopausal women [78]. Glutathione peroxidase (GPx) has important functions in the reduction of peroxides that are reported to inactivate vasodilating NO and the decomposition of S-nitrosoglutathione (GSNO), which plays an important role in vascular homeostasis [79]. Seleno-glutathione peroxidase mimic ebselen (PZ51)-protected endothelial and vascular system of spontaneous hypertensive rats (SHRsp) prone to stroke during pregnancy chronic hypertension [80]. It showed a significant relationship between low GPx level and vitiligo. Asian vitiligo patients showed lower levels of GPx than the controls, Caucasian populations, and healthy controls [81]. GPx polymorphism may contribute to the reduced GPx activity and the prevalence of vitiligo in Gujarat population [82]. In experimental models, GPx1 deficiency led to endothelial dysfunction, impaired angiogenesis, and increased infarction size and vascular permeability following ischemia/reperfusion injury [57, 83]. Furthermore, recent data [84] and others [85] showed that GPx1 deficiency provides atherosclerosis sensitivity in diabetic and hyperlipidemic environments, respectively. A recent study by Lubos et al. [86] showed that the reduction of GPx-1 activity in silenced human endothelial cells for GPx1 expression accelerated oxidative stress and increased nuclear factor kappa-light-chain-enhancer of activated B cells (NFκB) and c-Jun-N-terminal kinase (JNK) activation. GPx activity is associated with an increased risk of major adverse cardiovascular events at 1 year following an acute coronary syndrome. Serum GSHPx and arylesterase activity levels increased significantly after laryngectomy [87]. In a study [88], a decrease in GSH-Px activity was observed in patients with colon cancer. In another study [89], there was no difference in GPX or CAT activities in hypoxic and non-hypoxic patients. In one OSAS study [90], patients had lower GPX levels than healthy controls.

It is known that ROS have detrimental effects, which are controlled by various cellular defense system, consisting of enzymatic (catalase, glutathione peroxidase and superoxide dismutase, etc.) compounds [2, 91]. H_2O_2 is detoxified by catalase and glutathione peroxidase [22, 92]. Selenium is a trace element which is located in selenoenzymes, comprising thioredoxin reductase (TrxR) that is an enzyme which is related to the reduction of proteins and disulfides and glutathione peroxidase (GPx) for the detoxification of H_2O_2 [93–95].

Many studies show that the effect of organic and inorganic Se supplements on physiological functions and human health has been researched, but the most appropriate usage form of Se addition to dietary has not been specified [95–97].

2. Conclusion

The form of lipid peroxidation with the result of molecular oxygen conversion to ROS with various environmental factors particularly smoking cigarette, alcohol, UV rays, and other oxidants leads to oxidative stress. As a result of this, a multistage carcinogenesis process starts, and cells with diseases may occur by the degradation of the balance between both lipophilic and enzymatic antioxidants that form the antioxidant capacity of skin and ROS. Glutathione peroxidase activity is a primary antioxidant defense system that plays a key and fundamental role in the overall defense mechanisms and strategies in biological systems. There are at least eight GPx enzymes in humans, GPx1–GPx8. The GPx 1–8 genes are mapped to chromosomes 3, 14, 5, 19, 6, 6, 1, and 5, respectively. Glutathione peroxidases have also been implicated in the development and prevention of many common and complex diseases, including cancer and cardiovascular disease. Optimization of the nutritional status of selenium may result in higher GPx4 activity and thus delay or even prevent neuronal loss. Increasing selenium levels may reduce the risk of developing neurodegenerative disease in populations with low selenium exposure. The brain is vulnerable to increased ROS due to its high metabolic rate and relatively low antioxidant defense ability. Se deficiency has been associated with increased oxidative stress and neurodegenerative diseases. The role of Se proteins in the neurodegeneration of oxidative stress and ferropose can provide a unique insight into the mechanisms of cellular death in neurodegeneration. Optimization of nutritional status of Se may result in higher GPx activity. In a population with low selenium uptake, the toxic effects of mercury may be more pronounced because the metal forms an insoluble complex with selenium, thereby reducing bioavailability in various antioxidant systems (e.g., glutathione peroxidase). Recent developments in the new field of selenium biology and GPX1 have been shown to attempt to suggest the signaling and molecular mechanisms involved in glucose and lipid metabolism-related diseases. It mimics the requirements and opportunities of mimicking applications of various antioxidant enzymes in the treatment of insulin-dependent diseases. GPX enzyme is the main regulator of insulin physiology and energy metabolism. Hyperbaric oxygen, medical ozone, and enteral glutamine, alone or in combination with arginine, have shown positive effects on necrotizing enterocolitis (NEC) by modulating antioxidant defense mechanisms. Maintaining the physiological concentration of selenium is a prerequisite for preventing thyroid disease and maintaining general health. Supplementation with organic form is more effective and appears to be beneficial in immunological mechanisms in patients with autoimmune thyroiditis. Selenium supplementation has proven clinically beneficial in patients with mild to moderate Graves' orbitopathy. Decreased GPx activity causes hepatocellular degeneration and premature death of mice. Since many harmful conditions are known to directly disrupt the GPx, inhibition of ferroptosis may represent a suitable therapeutic approach to improve hepatocyte cell death. GPx is particularly vulnerable to the oxidative stress associated with hypertension. Low GPx levels may contribute to the pathogenesis of vitiligo in the Asian population as opposed to the White population. GPx1 plays a major role in vascular homeostasis. Specifically upregulating the activity of this isoform or designing functionally active mimetics may provide cardiovascular protection. GPx1 deficiency causes endothelial cell dysfunction and activation supporting atherogenesis. It can be concluded that serum GPx activity is low in patients compared to healthy control groups, and the balance between antioxidant and prooxidants is deteriorated in favor of prooxidants, and the deficiency of antioxidant enzymes in diseases may be a symptom in the explanation of cancer pathophysiology. The result could not be fully confirmed as there are some limitations. Limited numbers, small sample sizes, and methodological diversity may weaken statistical power. Higher quality studies with larger samples should be performed to confirm the results. Further studies are needed on this subject. We believe that this study will shed light on future studies.

Author details

Eren Sarıkaya[1]* and Selami Doğan[2]

1 Department of Chemistry, Division of Biochemistry, Faculty of Science, Yuzuncu Yil University, Van, Turkey

2 Bayrakli, Family Practice Center Number Four, Izmir, Turkey

*Address all correspondence to: erensaikaya1@gmail.com

References

[1] Arslan A, Demir H, Ozbay M, Arslan H. Evaluation of lipid peroxidation and some antioxidant activities in patients with primary and metastatic liver cancer. Journal of Cancer Therapy. 2014;**5**(2):192-197

[2] Gecit I, Aslan M, Gunes M, Pirincci N, Esen R, Demir H, et al. Serum prolidase activity, oxidative stress, and nitric oxide levels in patients with bladder cancer. Journal of Cancer Research and Clinical Oncology. 2012;**138**(5):739-743

[3] Altan N, Dinçel AS, Koca C. Diabetes mellitus ve oksidatif stres. Turkish Journal of Biochemistry. 2006;**31**(2):51-56

[4] Halliwell B. Drug antioxidant effects. Drugs. 1991;**42**(4):569-605

[5] Wei YH, Pang CY. The role of mitochondria in human aging process. Biotechnology International. 2005;**17**:8-13

[6] Gökpınar Ş, Koray T, Akçiçek E, Göksan T, Durmaz Y. Algal antioksidanlar. Journal of Fisheries and Aquatic Sciences. 2006;**23**(1/1):85-89

[7] Fuchs-Tarlovsky V. Role of antioxidants in cancer therapy. Nutrition. 2013;**29**(1):15-21

[8] Hristova M, Penev M. Oxidative stress and cardiovascular diseases. Trakia Journal of Sciences. 2014;**12**(3):296-303

[9] Derviş E. Oral antioksidanlar. SB. Haseki Eğitim Hastanesi, Dermatoloji Kliniği. Dermatoz Dergisi 2011;**2**(1):263-267

[10] Gülçin I. Isırgan Otunun (*Urtica dioica*) Antioksidant Aktivitesinin Belirlenmesi, Oksidatif Enzimlerin Karakterizasyonu ve Bazı In vivo Etkilerinin İncelenmesi (doktora tezi, basılmamış). AÜ, Fen Bilimleri Enstitüsü, Erzurum; 2002

[11] Lenehan P, Gutiérrez P, Wagner J, Milak N, Fisher G, Ross D. Resistance to oxidants associated with elevated catalase activity in HL-60 leukemia cells that overexpress multidrug-resistance protein does not contribute to the resistance to daunorubicin manifested by these cells. Cancer Chemotherapy and Pharmacology. 1995;**35**(5):377-386

[12] Glasauer A, Chandel N. Targeting antioxidants for cancer therapy. Biochemical Pharmacology. 2014;**92**(1):90-101

[13] Hayes J, McMahon M. NRF2 and KEAP1 mutations: Permanent activation of an adaptive response in cancer. Trends in Biochemical Sciences. 2009;**34**(4):176-188

[14] Mitsuishi Y, Taguchi K, Kawatani Y, Shibata T, Nukiwa T, Aburatani H, et al. Nrf2 redirects glucose and glutamine into anabolic pathways in metabolic reprogramming. Cancer Cell. 2012;**22**(1):66-79

[15] DeNicola G, Karreth F, Humpton T, Gopinathan A, Wei C, Frese K, et al. Oncogene-induced Nrf2 transcription promotes ROS detoxification and tumorigenesis. Nature. 2011;**475**(7354):106-109

[16] Hayes J, McMahon M, Chowdhry S, Dinkova-Kostova A. Cancer chemoprevention mechanisms mediated through the Keap1-Nrf2 pathway. Antioxidants & Redox Signaling. 2010;**13**(11):1713-1748

[17] Tai D, Jin W, Wu C, Si H, Cao X, Guo A, et al. Changes in intracellular redox status influence multidrug resistance in gastric adenocarcinoma

cells. Experimental and Therapeutic Medicine. 2012;**4**(2):291-296

[18] Aydin A, Arsova-Sarafinovska Z, Sayal A, Eken A, Erdem O, Erten K, et al. Oxidative stress and antioxidant status in non-metastatic prostate cancer and benign prostatic hyperplasia. Clinical Biochemistry. 2006;**39**(2):176-179

[19] Thapa D, Ghosh R. Antioxidants for prostate cancer chemoprevention: Challenges and opportunities. Biochemical Pharmacology. 2012;**83**(10):1319-1330

[20] De Bont R, Van Larebeke N. Endogenous DNA damage in humans: A review of quantitative data. Mutagenesis. 2004;**19**(3):169-185

[21] Aktaş M, Değirmenci U, Ercan SK, Tamer L, Atik U. Redükte glutatyon ölçümünde HPLC ve spektrofotometrik yöntemlerin karşılaştırılması. Türk Klinik Biyokimya Derg. 2005;**3**(3):95-99

[22] Matés JM, Sánchez-Jiménez F. Antioxidant enzymes and their implications in pathophysiologic processes. Frontiers in Bioscience. 1999;**4**(4):339-345

[23] Çaylak E. Hayvan ve bitkilerde oksidatif stres ile antioksidanlar. Tıp araştırmaları dergisi. 2011;**9**(1):73-83

[24] Fraternale A, Paoletti MF, Casabianca A, Nencioni L, Garaci E, Palamara AT, et al. GSH and analogs in antiviral therapy. Molecular Aspects of Medicine. 2009;**30**(1-2):99-110

[25] Pocernich CB, La Fontaine M, Butterfield DA. In-vivo glutathione elevation protects against hydroxyl free radical-induced protein oxidation in rat brain. Neurochemistry International. 2000;**36**(3):185-191

[26] Pompella A, Visvikis A, Paolicchi A, De Tata V, Casini AF. The changing faces of glutathione, a cellular protagonist. Biochemical Pharmacology. 2003;**66**(8):1499-1503

[27] Sies H. Glutathione and its role in cellular functions. Free Radical Biology and Medicine. 1999;**27**(9-10):916-921

[28] Wang J, Wang K, Wang Y, Lin S, Zhao P, Jones G. A novel application of pulsed electric field (PEF) processing for improving glutathione (GSH) antioxidant activity. Food Chemistry. 2014;**161**:361-366

[29] Reed DJ, Fariss MW. Glutathione depletion and susceptibility. Pharmacological Reviews. 1984;**36**(2):25S-33S

[30] Sozmen EY. Yaşlanma biyokimyası. In: Onat T, Emerk K, Sozmen EY, editors. İnsan Biyokimyası. Ankara: Palme Yayıncılık; 2002. pp. 665-674

[31] Esterbauer H, Gebicki J, Puhl H, Jürgens G. The role of lipid peroxidation and antioxidants in oxidative modification of LDL. Free Radical Biology and Medicine. 1992;**13**(4):341-390

[32] Arrick BA, Nathan CF. Glutathione metabolism as a determinant of therapeutic efficacy: A review. Cancer Research. 1984;**44**(10):4224-4232

[33] Akkuş İ. Serbest radikaller ve fizyopatolojik etkileri. Konya: Mimoza Yayınları; 1995

[34] Benzer F, Ozan ST. Fasciola hepatica ile enfekte koyunlarda lipid peroksidasyonu, antioksidant enzimler ve nitrik oksit düzeyleri. Turkish Journal of Veterinary and Animal Sciences. 2003;**27**:657-661

[35] Cheeseman KH, Slater TF. An introduction to free radical biochemistry. British Medical Bulletin. 1993;**49**(3):481-493

[36] Frei B. Reactive oxygen species and antioxidant vitamins: Mechanisms of action. The American Journal of Medicine. 1994;**97**(3):S5-S13

[37] Günaldı M. Kan selenyum düzeyi ve glutatyon peroksidaz aktivitesinin akut miyokart enfarktüsü gelişimi üzerine etkisi. Okmeydanı Eğitim ve Araştırma Hastanesi. İç Hastalıkları Kliniği Uzmanlık Tezi; 2009. 71p

[38] Öncü M, Gültekin F, Karaöz E, Altuntaş İ, Delibaş N. Klorprifos-Etil tarafından oluşturulan oksidatif hasarın sıçan karaciğerine etkileri. Turkiye Klinikleri Journal of Medical Sciences. 2002;**22**(1):50-55

[39] Çetin Ç. Koroner arter ektazi hastalarında oksidatif dna hasarı, lipid peroksidasyonu ve bazı antioksidant enzimler (yüksek lisans tezi, basılmamış), Yüzüncü Yıl Üniversitesi, Fen Bilimleri Enstitüsü, Kimya Anabilim Dalı; Van. 2011

[40] Yarıktaş M, Döner F, Doğru H, Aynalı G, Yönden Z, Delibaş N. Baş-boyun malign tümörlerinde malondialdehit düzeyleri ve antioksidan enzim aktiviteleri. SDÜ Tıp Fakültesi Dergisi. 2003;**10**(4). Available from: https://dergipark.org.tr/tr/pub/sdutfd/issue/20974/225466

[41] Cerutti P. Oxy-radicals and cancer. Lancet. 1994;**344**:862-863

[42] Çokal BG, Yurtdaş M, Güler SK, Güneş HN, Uçar CA, Aytaç B, et al. Serum glutathione peroxidase, xanthine oxidase, and superoxide dismutase activities and malondialdehyde levels in patients with Parkinson's disease. Neurological Sciences. 2017;**38**(3):425-431

[43] Ahmed RG. Gestational prooxidant-antioxidant imbalance may be at higher risk for postpartum thyroid disease. Endocrinology and Metabolic Syndrome. 2017;**6**(279):2161-1017

[44] Ruseva B, Atanasova M, Tsvetkova R, Betova T, Mollova M, Alexandrova M, et al. Effect of selenium supplementation on redox status of the aortic wall in young spontaneously hypertensive rats. Oxidative Medicine and Cellular Longevity. 2015:609053

[45] Giustarini D, Tsikas D, Colombo G, Milzani A, Dalle-Donne I, Fanti P, et al. Pitfalls in the analysis of the physiological antioxidant glutathione (GSH) and its disulfide (GSSG) in biological samples: An elephant in the room. Journal of Chromatography B. 2016;**1019**:21-28

[46] Ali MJ, Guesmi F, Harrath AH, Alwasel S, Hedfi A, Ncib S, et al. Investigation of antiulcer and antioxidant activity of *Juniperus phoenicea* L. (1753) essential oil in an experimental rat model. Biological and Pharmaceutical Bulletin. 2015;**38**(11):1738-1746

[47] Szczeklik K, Krzysciak W, Domagala-Rodacka R, Mach P, Darczuk D, Cibor D, et al. Alterations in glutathione peroxidase and superoxide dismutase activities in plasma and saliva in relation to disease activity in patients with Crohn's disease. Journal of Physiology and Pharmacology. 2016;**67**:709-715

[48] Moret-Tatay I, Iborra M, Cerrillo E, Tortosa L, Nos P, Beltrán B. Possible biomarkers in blood for Crohn's disease: Oxidative stress and microRNAs—Current evidences and further aspects to unravel. Oxidative Medicine and Cellular Longevity. 2016:2325162

[49] Crawford A, Fassett RG, Geraghty DP, Kunde DA, Ball MJ, Robertson IK, et al. Relationships between single nucleotide polymorphisms of antioxidant enzymes and disease. Gene. 2012;**501**(2):89-103

[50] Rosenberger A, Illig T, Korb K, Klopp N, Zietemann V, Wölke G, et al.

Do genetic factors protect for early onset lung cancer? A case control study before the age of 50 years. BMC Cancer. 2008;**8**(1):60

[51] Namikawa C, Shu-Ping Z, Vyselaar JR, Nozaki Y, Nemoto Y, Ono M, et al. Polymorphisms of microsomal triglyceride transfer protein gene and manganese superoxide dismutase gene in non-alcoholic steatohepatitis. Journal of Hepatology. 2004;**40**(5):781-786

[52] Nemoto M, Nishimura R, Sasaki T, Hiki Y, Miyashita Y, Nishioka M, et al. Genetic association of glutathione peroxidase-1 with coronary artery calcification in type 2 diabetes: A case control study with multi-slice computed tomography. Cardiovascular Diabetology. 2007;**6**(1):23

[53] Crawford A, Fassett RG, Coombes JS, Kunde DA, Ahuja KD, Robertson IK, et al. Glutathione peroxidase, superoxide dismutase and catalase genotypes and activities and the progression of chronic kidney disease. Nephrology, Dialysis, Transplantation. 2011;**26**(9):2806-2813

[54] Dutkiewicz G, Binczak-Kuleta A, Pawlik A, Safranow K, Wisniewska M, Ciechanowicz A, et al. Lack of association of C599T polymorphism in the glutathione peroxidase (GPX1) gene with delayed graft function, acute kidney graft rejection and chronic allograft nephropathy. Annals of Transplantation. 2010;**15**(3):30-34

[55] Chao CT, Chen YC, Chiang CK, Huang JW, Fang CC, Chang CC, et al. Interplay between superoxide dismutase, glutathione peroxidase, and peroxisome proliferator activated receptor gamma polymorphisms on the risk of end-stage renal disease among Han Chinese patients. Oxidative Medicine and Cellular Longevity. 2016:8516748

[56] Chabory E, Damon C, Lenoir A, Kauselmann G, Kern H, Zevnik B, et al. Epididymis seleno-independent glutathione peroxidase 5 maintains sperm DNA integrity in mice. The Journal of Clinical Investigation. 2009;**119**(7):2074-2085

[57] Forgione MA, Weiss N, Heydrick S, Cap A, Klings ES, Bierl C, et al. Cellular glutathione peroxidase deficiency and endothelial dysfunction. American Journal of Physiology. Heart and Circulatory Physiology. 2002;**282**(4):H1255-H1261

[58] Cardoso BR, Hare DJ, Bush AI, Roberts BR. Glutathione peroxidase 4: A new player in neurodegeneration? Molecular Psychiatry. 2017;**22**(3):328

[59] Cheng WH, Ho YS, Ross DA, Valentine BA, Combs GF Jr, Lei XG. Cellular glutathione peroxidase knockout mice express normal levels of selenium-dependent plasma and phospholipid hydroperoxide glutathione peroxidases in various tissues. The Journal of Nutrition. 1997;**127**(8):1445-1450

[60] Cho CS, Lee S, Lee GT, Woo HA, Choi EJ, Rhee SG. Irreversible inactivation of glutathione peroxidase 1 and reversible inactivation of peroxiredoxin II by H2O2 in red blood cells. Antioxidants & Redox Signaling. 2010;**12**(11):1235-1246

[61] Liddell JR, Hoepken HH, Crack PJ, Robinson SR, Dringen R. Glutathione peroxidase 1 and glutathione are required to protect mouse astrocytes from iron-mediated hydrogen peroxide toxicity. Journal of Neuroscience Research. 2006;**84**(3):578-586

[62] Wang X, Vatamaniuk MZ, Roneker CA, Pepper MP, Hu LG, Simmons RA, et al. Knockouts of SOD1 and GPX1 exert different impacts on murine islet function and pancreatic

integrity. Antioxidants & Redox Signaling. 2011;**14**(3):391-401

[63] Loh K, Deng H, Fukushima A, Cai X, Boivin B, Galic S, et al. Reactive oxygen species enhance insulin sensitivity. Cell Metabolism. 2009;**10**(4):260-272

[64] Ravn-Haren G, Olsen A, Tjønneland A, Dragsted LO, Nexø BA, Wallin H, et al. Associations between GPX1 Pro198Leu polymorphism, erythrocyte GPX activity, alcohol consumption and breast cancer risk in a prospective cohort study. Carcinogenesis. 2005;**27**(4):820-825

[65] Kuzuya M, Ando F, Iguchi A, Shimokata H. Glutathione peroxidase 1 Pro198Leu variant contributes to the metabolic syndrome in men in a large Japanese cohort. The American Journal of Clinical Nutrition. 2008;**87**(6):1939-1944

[66] Buraczynska M, Buraczynska K, Dragan M, Ksiazek A. Pro198Leu polymorphism in the glutathione peroxidase 1 gene contributes to diabetic peripheral neuropathy in type 2 diabetes patients. Neuromolecular Medicine. 2017;**19**(1):147-153

[67] Guerrero CH, Chávez PH, Castro NM, Carriedo AP, Del Río SG, Lizaur AP. Glutathione peroxidase-1 Pro200Leu polymorphism (rs1050450) is associated with morbid obesity independently of the presence of prediabetes or diabetes in women from Central Mexico. Nutrición Hospitalaria. 2015;**32**(4):1516-1525

[68] Shuvalova YA, Kaminnyi AI, Meshkov AN, Kukharchuk VV. Pro198Leu polymorphism of GPx-1 gene and activity of erythrocytic glutathione peroxidase and lipid peroxidation products. Bulletin of Experimental Biology and Medicine. 2010;**149**(6):743-745

[69] Nassi N, Ponziani V, Becatti M, Galvan P, Donzelli G. Anti-oxidant enzymes and related elements in term and preterm newborns. Pediatrics International. 2009;**51**(2):183-187

[70] Ozdemir R, Yurttutan S, Sarı FN, Uysal B, Unverdi HG, Canpolat FE, et al. Antioxidant effects of N-acetylcysteine in a neonatal rat model of necrotizing enterocolitis. Journal of Pediatric Surgery. 2012;**47**(9):1652-1657

[71] Piechota-Polanczyk A, Fichna J. The role of oxidative stress in pathogenesis and treatment of inflammatory bowel diseases. Naunyn-Schmiedeberg's Archives of Pharmacology. 2014;**387**(7):605-620

[72] Patlevič P, Vašková J, Švorc P Jr, Vaško L, Švorc P. Reactive oxygen species and antioxidant defense in human gastrointestinal diseases. Integrative Medicine Research. 2016;**5**(4):250-258

[73] Ventura M, Melo M, Carrilho F. Selenium and thyroid disease: From pathophysiology to treatment. International Journal of Endocrinology. 2017;(75):60

[74] Miranda-Díaz AG, Pazarín-Villaseñor L, Yanowsky-Escatell FG, Andrade-Sierra J. Oxidative stress in diabetic nephropathy with early chronic kidney disease. Journal Diabetes Research. 2016:7047238

[75] Gugliandolo A, Gangemi C, Calabrò C, Vecchio M, Di Mauro D, Renis M, et al. Assessment of glutathione peroxidase-1 polymorphisms, oxidative stress and DNA damage in sensitivity-related illnesses. Life Sciences. 2016;**145**:27-33

[76] Bednarek-Tupikowska G, Bohdanowicz-Pawlak A, Bidzińska B, Milewicz A, Antonowicz-Juchniewicz J, Andrzejak R. Serum lipid peroxide

levels and erythrocyte glutathione peroxidase and superoxide dismutase activity in premenopausal and postmenopausal women. Gynecological Endocrinology. 2001;**15**(4):298-303

[77] Serviddio G, Loverro G, Vicino M, Prigigallo F, Grattagliano I, Altomare E, et al. Modulation of endometrial redox balance during the menstrual cycle: Relation with sex hormones. The Journal of Clinical Endocrinology & Metabolism. 2002;**87**(6):2843-2848

[78] Klisic A, Kotur-Stevuljevic J, Kavaric N, Martinovic M, Matic M. The association between follicle stimulating hormone and glutathione peroxidase activity is dependent on abdominal obesity in postmenopausal women. Eating and Weight Disorders - Studies on Anorexia, Bulimia and Obesity. 2018;**23**(1):133-141

[79] Freedman JE, Frei B, Welch GN, Loscalzo J. Glutathione peroxidase potentiates the inhibition of platelet function by S-nitrosothiols. The Journal of Clinical Investigation. 1995;**96**(1):394-400

[80] Sui H, Wang W, Wang PH, Liu LS. Effect of glutathione peroxidase mimic ebselen (PZ51) on endothelium and vascular structure of stroke-prone spontaneously hypertensive rats. Blood Pressure. 2005;**14**(6):366-372

[81] Xiao BH, Shi M, Chen H, Cui S, Wu Y, Gao XH, et al. Glutathione peroxidase level in patients with vitiligo: A meta-analysis. BioMed Research International. 2016:3029810

[82] Laddha NC, Chatterjee S, Gani AR, Malek RA, Shah BJ, Begum R. Association of catalase T/C exon 9 and glutathione peroxidase codon 200 polymorphisms in relation to their activities and oxidative stress with vitiligo susceptibility in Gujarat population. Pigment Cell Research. 2007;**20**(5):405-407

[83] Forgione MA, Cap A, Liao R, Moldovan NI, Eberhardt RT, Lim CC, et al. Heterozygous cellular glutathione peroxidase deficiency in the mouse: Abnormalities in vascular and cardiac function and structure. Circulation. 2002;**106**(9):1154-1158

[84] Lewis P, Stefanovic N, Pete J, Calkin AC, Giunti S, Thallas-Bonke V, et al. Lack of the antioxidant enzyme glutathione peroxidase-1 accelerates atherosclerosis in diabetic apolipoprotein E-deficient mice. Circulation. 2007;**115**(16):2178

[85] Torzewski M, Ochsenhirt V, Kleschyov AL, Oelze M, Daiber A, Li H, et al. Deficiency of glutathione peroxidase-1 accelerates the progression of atherosclerosis in apolipoprotein E-deficient mice. Arteriosclerosis, Thrombosis, and Vascular Biology. 2007;**27**(4):850-857

[86] Lubos E, Kelly NJ, Oldebeken SR, Leopold JA, Zhang YY, Loscalzo J, et al. Glutathione peroxidase-1 deficiency augments proinflammatory cytokine-induced redox signaling and human endothelial cell activation. Journal of Biological Chemistry. 2011;**286**(41):35407-35417

[87] Bozan N, Demir H, Gürsoy T, Özkan H, Düzenli U, Sarıkaya E, et al. Alterations in oxidative stress markers in laryngeal carcinoma patients. Journal of the Chinese Medical Association. 2018;**81**(9):811-815

[88] Dusak A, Atasoy N, Demir H, Dogan E, Gursoy T, Sarikaya E. Investigation of levels of oxidative stress and antioxidant enzymes in colon cancers. Journal of Clinical and Analytical Medicine. 2017;**8**(6):469-473

[89] Wali SO, Bahammam AS, Massaeli H, Pierce GN, Iliskovic N, Singal PK, et al. Susceptibility of LDL

to oxidative stress in obstructive sleep apnea. Sleep. 1998;**21**(3):290-296

[90] Asker S, Asker M, Sarikaya E, Sunnetcioglu A, Aslan M, Demir H. Oxidative stress parameters and their correlation with clinical, metabolic and polysomnographic parameters in severe obstructive sleep apnea syndrome. International Journal of Clinical and Experimental Medicine. 2015;**8**(7):11449

[91] Matés JM, Pérez-Gómez C, De Castro IN. Antioxidant enzymes and human diseases. Clinical Biochemistry. 1999;**32**(8):595-603

[92] Hunt CR, Sim JE, Sullivan SJ, Featherstone T, Golden W, Von Kapp-Herr C, et al. Genomic instability and catalase gene amplification induced by chronic exposure to oxidative stress. Cancer Research. 1998;**58**(17):3986-3992

[93] Hail N Jr, Cortes M, Drake EN, Spallholz JE. Cancer chemoprevention: A radical perspective. Free Radical Biology and Medicine. 2008;**45**(2):97-110

[94] Arnér ES, Holmgren A. Physiological functions of thioredoxin and thioredoxin reductase. European Journal of Biochemistry. 2000;**267**(20):6102-6109

[95] Bennett LL, Rojas S, Seefeldt T. Role of antioxidants in the prevention of cancer. Journal of Experimental & Clinical Medicine. 2012;**4**(4):215-222

[96] Zhang W, Joseph E, Hitchcock C, DiSilvestro RA. Selenium glycinate supplementation increases blood glutathione peroxidase activities and decreases prostate-specific antigen readings in middle-aged US men. Nutrition Research. 2011;**31**(2):165-168

[97] Filippini T, Del Giovane C, Dennert G, Zwahlen M, Brinkman M, Zeegers MP, et al. Selenium for preventing cancer. The Cochrane Database of Systematic Reviews. 2018;**2018**(1)

Chapter 4

Neurodegenerative Diseases: Potential Effect of Glutathione

Aoula Moustapha

Abstract

Neurodegenerative diseases are characterized by the progressive deterioration of neuronal function in the central or peripheral nervous system and ultimately the death of nerve cells. There is a big evidence that oxidative stress is an essential mediator implicated in neurodegenerative processes and may be a key event triggering various forms of cell death. Here, we review the hypothesis that neuronal loss resulted from oxidative stress may be initiated by a drastic decrease in the antioxidant molecule glutathione (GSH). The impairment of physiological glutathione's levels and the alterations in the activities of its related enzymes in neuronal cells are increasingly suggested to be implicated in the initiation and progression of neurodegenerative diseases. GSH plays a vital role in cellular redox homeostasis in the nervous system and protects neurons against a variety of oxidative insults. GSH depletion can enhance oxidative stress and may increase protein aggregation leading to initiate cell death in distinct neuronal populations. Evidence demonstrates a grand impact of oxidative stress and loss neuronal GSH in Parkinson's disease and Alzheimer's disease.

Keywords: mitochondria, oxidative stress, protein aggregation, cysteine, redox signaling

1. Introduction

During the past thirty to forty years, and along with the global rise in life expectancy, neurodegenerative diseases of the brain that affect the elderly in particular have become a burden on society more and more. The European Union (EU) Joint Programme-Neurodegenerative Disease Research (JPND) states that by the year 2030, a quarter (25%) of the European population will be over age 65, a significant increase over the current 16% [1]. Thus, the scientists place a special focus on age-related neurodegenerative diseases (ADD) research. These ADD such as Alzheimer's Disease (AD) and Parkinson's Disease (PD) have become more common and have drawn a lot of attention due to their irreversibility and lack of effective treatment [2]. Neurodegenerative diseases (DD) are known by gradual damage in neural cells and neuronal loss, which conduct to impaired motor or cognitive function. Largely, treatment is accomplished by reducing symptoms more than researching disease physiology and heading to the mechanisms that limit disease progression [3]. The mediating pathogeneses in neurodegenerative diseases are still not fully illustrated; however, great evidence demonstrates that reactive oxygen species (ROS) could be a key event as an elevated level of oxidative stress (OS) has been observed in the brain of DD patients [4]. In recent decades, a broad range of studies has shown that the development of age-dependent neurodegeneration is due to a decrease in the antioxidant efficacy and an increase in oxidative damage. Cumulative ROS can cause damage to biomolecules such as lipids, proteins and DNA, in addition to mitochondrial dysfunction [5]. The increasing prevalence of DD and their profound hindrance to the quality of patient's life make it necessitous to come up with effective and novel treatment approaches, such as enhancing glutathione level in neurons. Glutathione (GSH) is a major antioxidant whose levels are found to decrease in aging as well as in DD. Scientists are currently heading to fully understand the role of GSH depletion in these diseases in addition to exploit that to develop GSH-based treatment. Glutathione is an essential cellular component, as it is primarily responsible for protecting and defending cells against any risk caused by exposure to ROS, and this role is evident, especially in the brain. Thus, GSH homeostasis disturbance and GSH-dependent enzymes inactivation lead to the breakdown of the main protective barrier against ROS and as a result, the cell becomes more vulnerable to the damage caused by OS [6].

In this chapter, we highlighted the OS in terms of its stimulating effect on the initiation and development of common ADD. We also demonstrated the great importance of glutathione in preserving nerve cells from the damage that OS may cause and the intracellular changes resulting from its depletion that may exacerbate the disease.

2. Oxidative stress and neurodegenerative disorders

2.1 Reactive oxygen species in brain

Although the human brain makes up only 2% of the body's weight, its oxygen requirements are estimated at 20% of the body's oxygen consumption. The brain is classified among the most generating organs for the reactive oxygen species (ROS), where 4% of the oxygen consumed by the mitochondria is converted into super oxide superoxide ion (O2•–), which possesses exceptionally high reactivity, particularly as a powerful oxidizing agent and an initiator of radical reactions. There are three radical reactions mainly initiated by superoxide "**Figure 1**" (i) under the influence of superoxide dismutase (SOD), the superoxide is converted into hydrogen peroxide (H_2O_2), which is subsequently converted to water and molecular oxygen by GSH peroxidase (GPx) or catalase. (ii) H_2O_2 can also react with iron, found in high concentration in brain, via the Fenton reaction to form hydroxyl radicals (OH.), which trigger lipid peroxidation. (iii) Superoxide also interacts with nitric oxide, which is formed in large quantities in the brain by an enzyme neuronal NO synthase (nNOS). This reaction is a million times faster than Fenton and produces the toxic oxidant peroxynitrite (ONOO−), which can spread 10,000 times farther than hydroxyl radicals. Harmful effects of ONOO− are varied including oxidation of macromolecules (DNA, proteins, lipids), nitration of amino acids and inactivating mitochondrial enzymes leading to mitochondrial dysfunction. It is now possible to detect specific markers that are byproducts of the oxidized endogenous macromolecules. For instance, 4-hydroxyl 2, 3-nenonal (HNE) is a marker of unsaturated lipid oxidation, which it has many cellular toxic effects such as irreversible formation of protein adduct and inhibition of GPx activity, and thus contributes in elevated levels of H_2O_2 [7].

2.2 Underlying role of oxidative stress in neurodegeneration

A large body of evidence demonstrates a particular susceptibility of neurons to ROS because of their distinctive characters: High energy demands, high oxygen

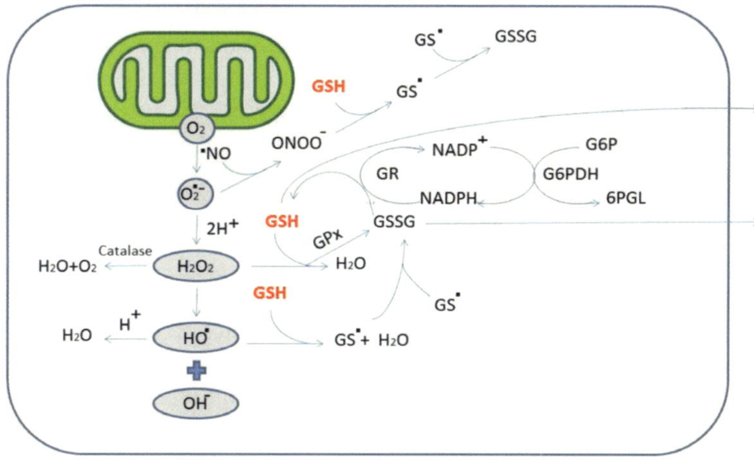

Figure 1.
Generation of ROS and Implication of glutathione in ROS/RNS elimination. As a result of mitochondrial respiration, the superoxide (O2-) is generated from O2. This latter can be converted into hydrogen peroxide (H_2O_2) by superoxide dismutase (SOD). A number of other ROS such as hydroxyl radicals (·OH) and hydroxyl anions (OH-) can be produced from H_2O_2. Hydroxyl radical and nitric oxide or peroxynitrite may interact directly with GSH forming GSSG. GSH serves as an electron donor for the reduction of H_2O_2 or other peroxides, catalyzed by GPx, and as result, it is converted to GSSG.

consumption, high levels of iron, polyunsaturated fatty acids and, in particular, low anti-oxidative protection [8]. The defense enzymatic system in neuronal cells is weak where the SOD, catalase, and GPx activities are low compared to other organs. In addition, glutathione, an essential anti-oxidant component, is present in the brain at low concentrations. These findings suggest the involvement of ROS in neurodegenerative diseases [4, 8, 9].

The OS was one of the important axes of research conducted to understand the pathogenesis of neurodegeneration. A number of research confirmed a strong involvement of oxidative stress in the pathophysiology of neurodegenerative diseases through a variety of mechanisms including induction of oxidation of macromolecules such as nucleic acids, proteins, lipids, mitochondrial dysfunction, glial cell activation, amyloid β deposition, apoptosis, and proteasome dysfunction [3, 5, 10, 11]. A systemic review showed that these mechanisms of neurodegeneration are involved in many harmful cellular pathways. It has been observed that the interference in these pathways in complex ways has the greatest impact on disease development [12], apoptosis, cytokine production and inflammatory responses, and proteasome dysfunction. Currently, there is an increasing focus on the effects of OS on the pathogenesis of neurodegenerative diseases (DD) and the effectiveness of antioxidants as a promising treatment for DD.

3. Glutathione homeostasis and neurodegenerative diseases

Glutathione (GSH) is a thiol-containing tripeptide of major significance in normal brain function. GSH is formed from glutamate, cysteine, and glycine. The γ-carboxyl group of glutamate links the N-terminal glutamate and cysteine residues, unusual peptide bond. This specific peptide bond protects GSH against cleavage by intracellular peptidases preventing its hydrolysis and making GSH moderately stable in the cell. In addition, the presence of the C-terminal glycine residue in GSH structure prevents its cleavage by intracellular γ-glutamyl cyclotransferase. The cysteine residue is an effective functional component of GSH as it provides thiol group, a principle responsible for the GSH activity. Moreover, cysteine residues form the intermolecular dipeptide bond in the oxidized form of GSH. Glutathione disulfide (GSSG), the major oxidized form, involves two residues of GSH that have been oxidized and connected by an intermolecular disulfide bond.

3.1 GSH content and synthesis

GSH is present in brain at elevated concentration (2–3 mM) compared to blood (15 μM) and cerebrospinal fluid (CSF) (5 μM) [13, 14]. Some evidence has been demonstrated that GSH is very poorly transported intact across the blood-brain barrier (BBB). However, it is probable that the blood is not the major source of cerebral GSH. This indicates that there is an avid brain system assures its synthesis in situ [15].

Generally, for maintaining GSH homeostasis in brain, there are at least two possible mechanisms: (i) glutathione constituents (cysteine moieties) may be recovered and recycled during the turnover of GSH in the brain, and (ii) precursors for brain glutathione synthesis (cysteine, cysteine-containing molecules) might be transported across the blood-brain barrier [13]. Cysteine is the rate-limiting substrate for neuronal GSH synthesis [15, 16]. In contrast, the availability of glutamine or glycine does not limit neuronal glutathione synthesis [13]. Therefore, cysteine alone is the crucial amino acid for neuronal GSH synthesis [17]. The neuronal uptake of cysteine is mediated by sodium-dependent systems, mainly the excitatory amino acid transporters (EAATs) [18]. EAATs have a significant function in removing extracellular glutamate in the CNS [19]. EAAT can transport not only excitatory amino acids, for example, glutamate and aspartate, but also cysteine, in particular, EAAT3, also known as EAAC1 that can transport cysteine at a rate comparable to that of glutamate [19].

Cystine, an oxidized form of two cysteines with a disulfide [20] linkage, is other source of free cysteine and employed as a substrate for GSH synthesis in some types of brain cells. Cystine moieties are transported into brain as (i) γ-glutamylcystine or as (ii) cystinylbisglycine which are possible origins of GSH in brain [21]. Cystine is especially important in maintaining glutathione levels in astrocytes [22], while it has no significance in the synthesis of neuronal GSH due to the inability of neurons to uptake it. Therefore, Content of cysteine or cysteine precursors determines the glutathione level in neurons since neurons are not able to use the cystine but rather rely on the availability of cysteine for their glutathione synthesis [20]. In addition to cysteine, neurons can utilize the cysteine donors such as CysGly, γGluCys, and N-acetylcystein (NAC) as precursors for glutathione. The presence of methionine does not increase neuronal glutathione levels [23]. Methionine is the main precursor of cysteine in liver, which supplies 50% of the cysteine needed for GSH synthesis. However, its role in producing cysteine in the brain is negligible and thus the neuronal GSH synthesis is not related to supply of methionine [16]. Among the exogenous precursors of glutathione, the dipeptide CysGly may be the most important. CysGly is efficiently utilized by neurons in micromolar concentrations [24].

Astrocytes store and synthesize high levels of GSH compared to neurons [13, 25, 26]. This is explained by the inability of neurons to directly uptake GSH. As well as, neurons utilize cysteine, not cystine, for GSH synthesis, whereas astrocytes utilize both [27, 28]. According to the above, neurons rely mainly on astrocytes to supply the necessary cysteine to neuronal GSH synthesis. GSH, released by astrocytes, undergoes a cleavage process by γ-glutamyl transpeptidase (γGT) [29] producing a γ-glutamyl moiety and a dipeptide CysGly which is an essential precursor of neuronal glutathione. The dipeptide CysGly could be uptake into neurons via a peptide transporter as has been described for astrocytes [30]. The dipeptide CysGly is hydrolyzed, upon entry into the neuron, by a neuronal ectopeptidase, providing cysteine and glycine [20] which subsequently are taken up as precursors for glutathione synthesis. Glutathione is synthesized by two successive enzymatic steps dependant on ATP [13, 20]. the first step include γ-glutamylcysteine synthetase (GCL) which mediates the first reaction between glutamate and cysteine to form a dipeptide, γ-glutamylcysteine (γGluCys) which in turn combines with glycine to produce GSH. When a sufficient amount of glutathione is synthesized, a feedback occurs where GCL is inhibited [31]. Conversely, GSH depletion causes in the short term an increase in GCL activity and consequently an increase in GSH synthesis.

3.2 GSH activity against OS

The adult mammalian brain has a great demand for energy, and it almost relies entirely on metabolism of glucose. Most of the glucose is completely oxidized to carbon dioxide to meet energy requirements. This very high ability to oxidize glucose indicates that the brain may produce ROS at a remarkable rate. This increase in ROS production combines with low levels of defense mechanisms such as catalase and a high lipid content in brain. All of these indicate that the brain may be particularly vulnerable to OS.

GSH plays a leading role in reducing high levels of ROS and minimizing oxidative damage in brain (**Figure 1**). This importance has been established by several studies demonstrating that OS was aggravated by the GSH depletion, while increased intracellular GSH improved this damage [32]. GSH is a great component that provides protection against OS in brain by a direct interact with superoxide [33], NO [34], hydroxyl radical [35], and ONOO– [36]. The GSH capacity to scavenge superoxide is higher than NAC or cysteine [37]. Moreover, GSH is a principle hydroxyl radical scavenger because of unavailability of enzymatic defense against these radicals. On the other hand, GSH participates in enzyme-catalyzed redox cycling. The most important enzyme in glutathione redox reaction is glutathione peroxidase (GPx) due to its leading role in the reduction of toxic H_2O_2 (or lipid peroxide, ROOH) to H_2O (or ROH). GSH serves as an electron donor for the reduction of H_2O_2 or other peroxides, catalyzed by GPx, and as result, it is converted to GSSG [21]. The glutathione redox cycle, is completed by glutathione reductase (GR). This GSH redox cycle takes place in the cytosol and mitochondria, whereas GSH is compartmentalized in mitochondria [38], the major intracellular source of ROS [39], after its synthesis in cytosol. Catalase also reduces H_2O_2 to H_2O but it is unable to detoxify lipid peroxides and is not exist in mitochondria of most tissues. For these reasons, GPx is especially significant in protecting of mitochondria against H_2O_2,that are constantly generated during cell respiration [40, 41]. Mitochondria contain 5–15% of total cellular GSH [42]. The maintenance of this mitochondrial GSH pool occurs through the action of a high-affinity GSH uptake system [43] which is a main determinant of neuronal susceptibility to OS [44]. Depletion of this pool in brain mitochondria makes them more vulnerable to toxic effects of H_2O_2 leading to irreversible damages [45] and death. If mitochondria are not protected against OS insult, the organelles become irreversibly damaged through a process culminating with induction of a mitochondrial permeability transition (mPT) which is associated with the collapse of mitochondrial membrane ($\Delta\Psi$) and colloid osmotic swelling of the matrix [46]. As well, GSH detoxifies many agents that can induce the mPT in brain mitochondria including 4-hydroxyhexenal (a lipid peroxide) [39]. These findings indicate that GSH has a high significance in maintaining mitochondrial integrity in brain and other organs. Moreover, GSH is a substrate for glutathione S-transferase (GST) that catalyze GSH-dependent reduction of lipid peroxides. In addition to the above, there is a potential synergistic relationship between reduced glutathione and vitamin E, another line of defense. This vitamin is well recognized as antioxidant incorporating into cellular membranes to inhibit lipid peroxidation [47]. Lipids are protected against ROS by α-tocopherol (vitamin E), which quenches ROS and by that, converts to α-tocopheroxyl radical. This latter can re-reduced non-enzymatically to α-tocopherol by GSH [48]. This reaction and those that are catalyzed by GPx and GST possess peroxidase activity and form a protective barrier of the brain against damaging effects of H_2O_2 on polyunsaturated fatty acids in biomembranes (lipid peroxidation) [49].

Many of studies have been demonstrated the specific toxicity of Hydrogen peroxide to brain [42, 43, 50]. This peroxide induces apoptosis in neuronal cells which are particularly sensitive to its toxic effects [51]. Nevertheless, neurons can detoxify H_2O_2, but apparently this capacity is more greater in astrocytes for which they play a putative role in the modulation of the neurotoxic effects of H_2O_2 [45, 46, 52]. The neuronal defense system against H_2O_2 is mainly based on glutathione redox cycle. This role of GSH is clearly illustrated by a rapid oxidation of GSH when H_2O_2 is applicated to neurons [53]. Intracellular GSH depletion enforces mitochondrial damage and causes cell death. Apoptosis has been hypothesized to be mediated through the induction of free radicals via oxidative pathways. Thus, a direct cause/effect relationship between GSH depletion and apoptosis was evidenced in neuronal cell [54]. In addition, GSH depletion is an early hallmark in the progression of cell death [55].

3.3 Implication of glutathione in neurodegenerative disorders

It has been previously emphasized that the breakdown of the balance between ROS and antioxidant defense systems is the main manipulator triggering the initiation or progression of a number of common neurodegenerative diseases such as Parkinson's (PD) and Alzheimer's (AD) diseases. Each of these diseases depends on a number of factors including mainly OS. However, the causative link between OS and neurodegeneration is not in the scope of this part as it focuses on the dysregulation of the GSH-based antioxidant network in the context of common neurodegenerative diseases: Parkinson's disease, Alzheimer's disease [6].

3.3.1 Parkinson's disease (PD)

The primary pathologic hallmarks of PD are loss of dopaminergic neurons located in an area of the brain called the substantia nigra pars compacta, and the presence of Lewy bodies, intracellular aggregates of misfolded α-synuclein, in dopaminergic neurons and likely contribute to the death of these neurons. Neurons in the substantia nigra pars compacta produce dopamine, a neurotransmitter (chemical messenger) that transmits signals from the substantia nigra to other parts of the brain. These other parts of the brain are collectively called the "basal ganglia". Communication among neurons of the substantia nigra pars compacta and the basal ganglia produce smooth, purposeful movement. When the neurons in the substantia nigra are damaged in large numbers, the loss of dopamine prevents normal function in basal ganglia and causes the motor symptoms of PD: tremor, rigidity, impaired balance, and loss of spontaneous movement [56].

Dopaminergic SN cells are usually pigmented with black neuromelanin, produced from of the autoxidation [57] or enzyme-mediated oxidation [58] of the cytoplasmic dopamine (DA) to DA-o-quinone, which then Polymerizes. Usually, this process is accompanied with production of H_2O_2 rendering dopaminergic SN cells are particularly sensitive to OS probably. It has been reported that dopaminergic SN neurons having high basal levels of DA oxidation, heavily pigmented, is particularly vulnerable to degeneration in PD [59].

A massive loss of nigral GSH is the most notable distinctive changes that occur in the earliest stage [60] in the parkinsonian SN. This GSH loss is uncorrelated to altered activities of biosynthetic enzyme and not accompanied by an increase in GSSG levels [61]. It has been indicated that the drastic drop in GSH is attributed to raise in activity of γ-GT, causing an increased removal of both GSH and GSSG from cells [61]. It is interesting that GSH depletion is characteristic to the parkinsonian SA and is not observed in other neurodegenerative disorders of the basal ganglia [62].

GSH depletion cause indirectly formation of endotoxins in the cytoplasm of pigmented SN cells that contribute to the degeneration of these neurons in PD "**Figure 2**." As previously mentioned the activity of γ-GT is up-regulated significantly in the parkinsonian SN [61]. This enzyme is involved in translocation of free cysteine into dopaminergic SN neurons and expulsion of GSH out of these cells. Thus, the profound loss of nigral GSH, main storage form of cysteine, makes the free cysteine, which is increasingly transported into the cell, more likely to bind to oxidizing dopamine and formation DHBTs (dihydrobenzothiazines) by a series of consecutive reactions [63]. These compounds are lethal and evoke profound neurobehavioral responses, especially DHBT-1 which cause irreversible inhibition of mitochondrial complex I [64].

The presence of Lewy bodies, aggregated misfolded α-synuclein, in SN is also characteristic hallmarks to PD which mainly participate in neurodegeneration [65]. The plurality of intracellular proteins is degraded by the ubiquitin (Ub)–proteasome pathway (UPP). In this pathway, the protein Ub, belongs to a family of heat shock proteins (HSPs), is covalently bound by thiol groups to misfolded or damaged proteins and contributes in their breakdown by transferring them to the protease 26S complex. There are three enzymes contributing to UPP: E1 (Ub-activating enzyme) and E2s (Ub-carrier) prepare Ub for conjugation, but the main enzyme in the process is the E3 (Ub-protein ligase) which transfers the activated ubiquitin to the protein substrate to be degraded [66]. Any defect in the components of UPP or a lack of their activity result in accumulation of α-synuclein protein and subsequent aggregation leading in turn to the formation of Lewy bodies. The depletion in GSH in dopaminergic SN neurons leads to decreased E1 activity and subsequent UPP

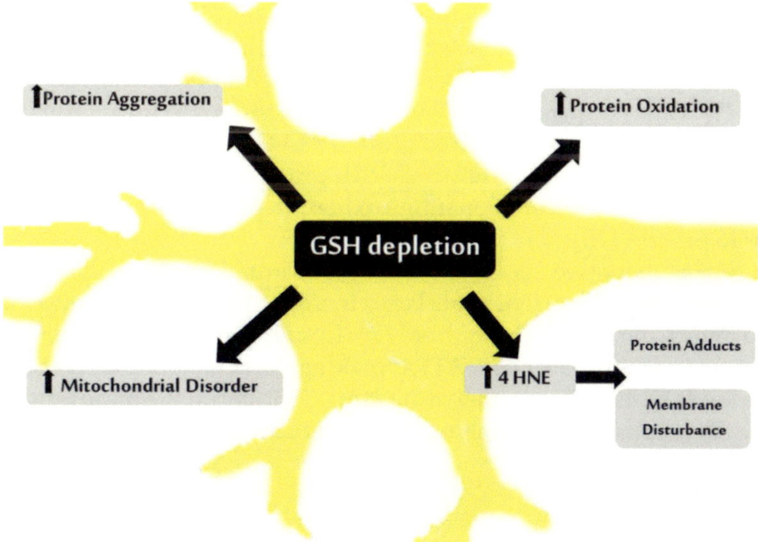

Figure 2.
Consequences of GSH depletion in SN dopaminergic neuronal cells in PD. Drastic loss of GSH is associated with protein aggregation which form Lewy bodies, mitochondrial dysfunction resulted from inhibition of complex I activity and oxidative damage including protein oxidation and the deleterious effects of the lipid peroxidation by-product 4HNE.

disturbance [67]. This finding indicates that GSH protects the active sites of these enzymes from being oxidized during oxidative stress, and thus keeps them performing in the Ub-proteasome pathway.

Additionally, the early GSH loss in parkinsonian SN accompanied by increased OS leads to raise in oxidized proteins. In the early stage of PD, HSP proteins are expressed at high levels to prevent the deleterious effects resulting from accumulation and aggregation of damaged proteins in dopaminergic neurons. As the disease progresses, these defenses become unable to control the build-up of protein aggregates [68].

OS also target the mitochondria and interfere with all of their functions. Mitochondrial disorders occupy a crucial place in the mechanisms that mediate neurodegeneration associating with the pathology of PD [69]. Since glutathione is the main component in detoxification of hydroperoxides in mitochondria, its depletion in the brain is believed to promote mitochondrial insult most likely via increased ROS. The mitochondria are known by their vulnerability to OS that might interfere with all of their functions. By serving as the main component in detoxification of hydroperoxides in mitochondria, GSH may reduce the oxidative insults that affect mitochondria. GSH depletion in the brain therefore is believed to promote mitochondrial insult most likely via increased ROS [70].

In synaptic mitochondria, the major role in control over oxidative phosphorylation is attributed to complex I that at 25% inhibition, energy metabolism is disturbed and ATP synthesis is drastically affected. To manifest similar effects [71], complex III and IV inhibition up to 80% is necessitated. The reduced complex I activity in the SN is known as a considerable biochemical characteristic in Parkinsonian brain [72]. Evidence suggest that Early depletions in nigral GSH levels may be directly lead to mitochondrial complex I activity inhibition and subsequent mitochondrial dysfunction which ultimately induces dopaminergic cell death related to PD. The complex I is the most severely influenced mitochondrial enzymes during OS [73]. It is believed that OS, due to decreased GSH availability in the brain, is the major responsible of mitochondrial complex I activity inhibition. This susceptibility of complex I to OS might be explicated by the oxidation of thiol (SH) groups of protein and the existence of accessible oxidation sensitive iron-sulfur centers in this complex [74].

It is recognized that GSH controls the activity of thiol-dependent proteins by keeping the SH groups of protein in a reduced state and preventing them from oxidation [75]. GSH conjugates with oxidized thiol groups to form protein-SS-G and subsequently can be re-reduced to protein and GSH by GR, thioredoxin or protein disulfide isomerase. In addition, GSH, present in dopaminergic cells, can bind to quinones resulted from dopamine oxidation preventing their reaction with SH groups in protein [76].

Lastly, during oxidative insult, aldehydes are formed as a byproduct; the most common of these types is 4HNE. This latter is able to incorporating into the membranes causing changes in their fluidity [77]. In addition, 4HNE can form adducts with important proteins like Na/K ATPas making them inactive. GSH may help reduce the levels of 4HNE by conjugating it via GST. In the PD brains, the loss in GSH in the SN results in high levels of 4HNE adducts [78].

3.3.2 Alzheimer's disease (AD)

AD, the most common age-related neurodegenerative disease, is known by progressive dementia affecting older populations. This disease is pathologically characterized by depositions of amyloid β (Aβ) plaques and neurofibrillary tangles (NFTs) [14]. The presence of amyloid plaques, which are mainly composed of Aβ peptide, in the extracellular space of AD brain is a main hallmark of disease. The excess of

Aβ levels, especially Aβ 42, the most neurotoxic peptide, causes the emergence of familial forms of Alzheimer's disease. This increase in Aβ 42 leads to the formation of soluble oligomers, causing permanent changes in synaptic function. In parallel, Aβ 42 is aggregated forming mostly β-sheet rich fibrils that enhance local inflammatory responses (microgliosis and astrocytosis). Synaptic spine loss and neurotic dystrophy are also observed. Over time, these events result in a biochemical changes including oxidative stress, altered ionic (e.g.; calcium) homeostasis [79]. Amyloid plaques are the determining factor in triggering a signaling pathway leading to AD progression. Recent evidence suggests that Aβ plaques induce neuronal apoptosis in the brain and in primary neuronal cultures, and this Aβ-induced neuronal death may be responsible in part for the cognitive decline found in AD patients. In addition, aggregated Amyloid-β activates the p38 mitogen activated protein kinase (MAPK) in cell leading to hyperphosphorylation of protein Tau and formation of neurofibrillary tangles (NFTs) inside neurons, making the microtubules unstable and causing the loss of neuron functionality [80].

Evidence demonstrates that Soluble Aβ oligomers are able to block the EAAC1-mediated cysteine uptake leading to a GSH loss in cultured human neuronal cells [81]. This is supported by autopsy brain of AD patients, which exhibit aberrant EAAC1 accumulation in pyramidal neurons of the hippocampus [82] and decreased GSH/GSSG ratios with the progress of disease [83].

Based on the above, it is possible to emphasize the notion of EAAC1 dysfunction in Alzheimer's disease.

Oxidative stress is considered a major pathogenic factor in AD. Since GSH depletion is of immense implication in oxidative stress, it is expected to have a role in the emergence and development of the disease. A recent clinical study using NMR spectroscopy showed that GSH level is depleted in AD patients as compared to healthy subjects [84]. This finding may have a profound clinical significance. In addition, the analysis of the blood samples of AD patients showed a decrease of GSH concentration in red blood cells compared to age- and gender-matched controls [85]. This is also observed in mild cognitive impairment (MCI) which is a preclinical stage of AD. MCI patients showed a decrease in GSH/GSSG ratios and GST activity in the hippocampus compared to healthy age-matched controls [86]. According to these results, it is suggested that disturbances in GSH metabolism precede the onset of AD. Genetic polymorphisms in the GPx-1 and GST genes were identified as positive risk factors for AD [87]. This can be the reason of decreases in GPx and GST activities in AD [88].

As was previously mentioned, ROS formation is induced by Aβ aggregating and cause a number of oxidative damages and metabolic insults including generation of HNE in hippocampal neurons, which could in turn mediate the toxicities of such insults [89]. Several studies have been shown an increase in lipid peroxidation in the brain of AD patients compared with age-matched controls [90]. As a result of lipid peroxidation, HNE, secondary bioactive aldehyde, is produced at a high levels in several brain regions of late-stage AD subjects [89]. The significant role of HNE in the progression of AD is supported by many findings. Accordingly, an increased level protein-bound HNE in brain of MCI patients was observed [91]. Many proteins were found to be significantly HNE-modified in AD such as ATP synthase, glutamine synthase, DRP-2, and MnSOD. These proteins have a great implication in the regulation of structural functions of brain cell in addition to a number of cellular functions including cellular signaling, energy metabolism and detoxification. Evidence showed that GSH could prevent oxidative damage induced by Aβ and HNE in cultured neuronal cells. This finding suggests that GSH depletion exacerbates oxidative insults stimulated by Aβ and HNE and therefore accelerates the development of the disease.

4. Conclusion

GSH is an interesting subject studied intensively in the brain for the past several decades. The purpose of such research is not only to understanding the potential role of intracellular GSH in preventing DD progression but also to provide the mechanistic insights contributing to the cellular dysfunctions associated with these diseases. GSH depletion is a common feature of DD triggered by a wide variety of cause including disturbance in GSH homeostasis and modification of the GSH related enzymes. Multiple cellular problems attributed to dysregulation of GSH and GSH-dependent enzymes contributes to impairment in the function of mitochondria, elevation in oxidative damage, disruption of intracellular signal transduction pathways, protein aggregation, and eventually cell death.

It is important to note that further research is necessary to determine more accurately the involvement of disruption of the network of glutathione-dependent reactions in the neurodegenerative events and find new ways to prevent or limit these events. As well to suggest more effective approaches therapy for DD patients.

Acknowledgements

The author would like to thank team of faculty of pharmacy Al-Baath University, and IntechOpen team.

Acronyms and abbreviations

EU	European union
JPND	joint programme-neurodegenerative disease research
GSSG	glutathione disulfide
GPx	glutathione peroxidase
GST	glutathione S-transferase
GS	glutathione synthetase
GR	glutathione reductase
PD	Parkinson's disease
AD	Alzheimer's Disease
DD	neurodegenerative diseases
RNS	reactive nitrogen species
ROS	reactive oxygen species
GSH	reduced glutathione
OS	oxidative stress
4HNE	4-hydroxyl 2, 3-nenonal
EAAC1	excitatory amino acid transporter C1
EAAT	excitatory amino acid transporter
GCS	γ-glutamylcysteine synthetase
NFTs	neurofibrillary tangles

Author details

Aoula Moustapha
Head of Pharmacology and Toxicology Department, Faculty of Pharmacy, Al-Baath University, Homs, Syria

*Address all correspondence to: omoustapha@albaath-univ.edu.sy; aoulamou@yahoo.com

References

[1] Rekatsina M. Pathophysiology and therapeutic perspectives of oxidative stress and neurodegenerative diseases: A narrative review. Advances in Therapy. 2020;**37**:113-139. DOI: 10.1007/s12325-019-01148-5

[2] Piccardi L, Curcio G, Palermo L, Pai M. Ageing and neurodegenerative disorders. Behavioural Neurology. 2015;**2015**:2-4

[3] Liu Z, Zhou T, Ziegler AC, Dimitrion P, Zuo L. Oxidative stress in neurodegenerative diseases: From molecular mechanisms to clinical applications. Oxidative Medicine and Cellular Longevity. 2017;**4**:1-11

[4] Niedzielska E, Smaga I, Gawlik M, Moniczewski A. Oxidative stress in neurodegenerative diseases. Molecular Neurobiology. 2016;**53**:4094-4125. DOI: 10.1007/s12035-015-9337-5

[5] Gandhi S, Abramov AY. Mechanism of oxidative stress in neurodegeneration. Oxidative Medicine and Cellular Longevity. 2012;**2012**:1-111. DOI: 10.1155/2012/428010

[6] Johnson WM, Wilson-delfosse AL, Mieyal JJ. Dysregulation of glutathione homeostasis in neurodegenerative diseases. Nutrition. 2012;**4**:1399-1440. DOI: 10.3390/nu4101399

[7] Adam-Vizi V. Production of reactive oxygen species in brain mitochondria: Contribution by electron transport chain and non-electron transport chain sources. Antioxidants & Redox Signaling. 2005;**7**:1140-1149

[8] Salim S. Oxidative stress and the central nervous system. The Journal of Pharmacology and Experimental Therapeutics. 2017;**360**:201-205

[9] Kim GH, Kim JE, Rhie SJ, Yoon S. The role of oxidative stress in neurodegenerative diseases. Experimental Neurology. 2015;**24**(4): 325-340

[10] Chen X, Guo C, Kong J. Oxidative stress in neurodegenerative diseases. Neural Regeneration Research. 2012;**7**(5):376-385. DOI: 10.3969/j.issn.1673-5374.2012.05.009

[11] Butterfield DA, Reed T, Newman SF, Sultana R. Roles of amyloid β-peptide-associated oxidative stress and brain protein modifications in the pathogenesis of Alzheimer's disease and mild cognitive impairment. Free Radical Biology & Medicine. 2007;**43**:658-677. DOI: 10.1016/j.freeradbiomed.2007.05.037

[12] Yaribeygi SA, Panahi Y, Javadi B. The underlying role of oxidative stress in neurodegeneration: A mechanistic review. CNS & Neurological Disorders Drug Targets. 2018;**17**(3):2018

[13] Dringen R. Metabolism and functions of glutathione in brain. Progress in Neurobiology. 2000;**62**:649-671. DOI: 10.1016/j.pneurobio.2014.01.002

[14] Aoyama K, Nakaki T. Impaired glutathione synthesis in neurodegeneration. International Journal of Molecular Sciences. 2013;**14**(10):21021-21044. DOI: 10.3390/ijms141021021

[15] Ballatori N, Krance SM, Notenboom S, Shi S, Tieu K, Hammond CL. Glutathione dysregulation and the etiology and progression of human diseases. Biological Chemistry. 2009;**390**(3): 191-214. DOI: 10.1515/BC.2009.033. Glutathione

[16] Aoyama K, Watabe M, Nakaki T. Regulation of neuronal glutathione synthesis. Journal of Pharmacological Sciences. 2008;**108**(3):227-238. DOI: 10.1254/jphs.08R01CR

[17] Dringen R, Pfeiffer B, Hamprecht B. Synthesis of the antioxidant glutathione in neurons: Supply by astrocytes of CysGly as precursor for neuronal glutathione. The Journal of Neuroscience. 1999;**19**(2):562-569. DOI: 10.1523/jneurosci.19-02-00562.1999

[18] Shanker G, Allen JW, Mutkus LA, Aschner M. The uptake of cysteine in cultured primary astrocytes and neurons. Brain Research. 2001;**902**: 156-163

[19] Magi S, Piccirillo S, Amoroso S, Lariccia V. Excitatory amino acid transporters (EAATs): Glutamate transport and beyond. International Journal of Molecular Sciences. 2019;**20**: 5674

[20] Dringen R, Gutterer JM, Gros C, Hirrlinger J. Aminopeptidase N mediates the utilization of the GSH precursor CysGly by cultured neurons. Journal of Neuroscience Research. 2001;**66**(5):1003-1008. DOI: 10.1002/jnr.10042

[21] Cooper KB. Multiple roles of glutathione in the central nervous system. Biological Chemistry. 1997;**378**: 793-802

[22] Stephanie HS, Roberta SM, Ogunrinu-Babarindea T, Holta KT. Role of glutamate transporters in redox homeostasis of the brain. Neurochemistry International. 2014;**0**:181-191. DOI: 10.1016/j.neuint.2014.01.001.Role

[23] Dringen R, Hamprecht B. N-Acetylcysteine, but not methionine or 2-oxothiazolidine-4- carboxylate, serves as cysteine donor for the synthesis of glutathione in cultured neurons derived from embryonal rat brain. Neuroscience Letters. 1999;**259**:79-82

[24] Dringen R, Gutterer JM, Hirrlinger J. Glutathione metabolism in brain metabolic interaction between astrocytes and neurons in the defense against reactive oxygen species. 2000;**4916**:4912-4916

[25] Dringen R, Kranich O, Hamprecht B. Use of dipeptides for the synthesis of glutathione by astroglia-rich primary cultures. Journal of Neurochemistry. 1997;**69**:868-874

[26] Dringen R. Glutathione restoration as indicator for cellular metabolism of astroglial cells. Developmental Neuroscience. 1998;**20**:401-407

[27] Kranich O, Hamprecht B, Dringen R. Different preferences in the utilization of amino acids for glutathione. Neuroscience Letters. 1996;**219**:211-214

[28] Sagara JI, Miura K, Bannai S. Maintenance of neuronal glutathione by glial cells. Journal of Neurochemistry. 1993;**61**(5):1672-1676. DOI: 10.1111/j.1471-4159.1993.tb09802.x

[29] Dringen R, Kranich O. The gamma-glutamyl transpeptidase inhibitor acivicin preserves glutathione released by astroglial cells in culture. Neurochemical Research. 1997;**22**(6):727-733

[30] Dringen R, Hamprecht B, Bröer S. The peptide transporter PepT2 mediates the uptake of the glutathione precursor CysGly in astroglia-rich primary cultures. Neurochemistry. 1998;**71**:388-393

[31] Richman G. Regulation feedback of y-glutamyl-cysteine synthetase inhibition by glutathione. The Journal of Biological Chemistry. 1974;**250**(4):1422-1427

[32] Kwon DH et al. Protective effect of glutathione against oxidative macrophages through activating the nuclear factor oxygenase-1 pathway.

Antioxidants. 2019;**8**:82. DOI: 10.3390/antiox8040082

[33] Winterbourn Christine C, Metodiewa D. The reaction of superoxide with reduced glutathione. Archives of Biochemistry and Biophysics. 1994;**314**:284-290

[34] Clancy RM, Levartovsky D, Leszczynska-piziak J, Yegudin J, Abramson SB. Nitric oxide reacts with intracellular glutathione and activates the for S-nitrosoglutathione as a bioactive intermediary hexose monophosphate shunt in human neutrophils: Evidence. Proceedings of the National Academy of Sciences of the United States of America. 1994;**91**: 3680-3684

[35] Bains JS, Shaw CA. Neurodegenerative disorders in humans: The role of glutathione in oxidative stress-mediated neuronal death. Brain Research Reviews. 1997;**25**:335-358

[36] Barker JE, Bolafiosb JP, Land JM, Clark JB. Glutathione protects astrocytes from peroxynitrite-mediated mitochondrial damage: Implications for neuronal/astrocytic trafficking and neurodegeneration. Developmental Neuroscience. 1996;**18**:391-396

[37] Hussain S, Jr S, Ali SF. Role of metallothionein and other antioxidants in scavenging superoxide radicals and their possible role in neuroprotection. Neurochemistry International. 1996;**29**(95):145-152

[38] Ribas V, García-Ruiz C. Glutathione and mitochondria. Frontiers in Pharmacology. 2014;**5**:151. DOI: 10.3389/fphar.2014.00151

[39] Starkov AA. The role of mitochondria in reactive oxygen species metabolism and signaling. Annals of the New York Academy of Sciences. 2008;**1147**:37-52. DOI: 10.1196/annals.1427.015.The

[40] Calabrese G, Morgan B. Mitochondrial glutathione: Regulation and functions. Antioxidants & Redox Signaling. 2017;**27**(15):1162-1177. DOI: 10.1089/ars.2017.7121

[41] Benzi G, Moretri A. Age- and peroxidative stress-related modifications of the cerebral enzymatic activities linked to mitochondria and the glutathione system. Free Radical Biology & Medicine. 1995;**19**(1):77-101

[42] Meredith MJ, Reed DJ. The mitochondrial pool of glutathione in the isolated hepatocyte. The Journal of Biological Chemistry. 1982;**257**(7):3747-3753

[43] Johannes Martensson AM, Lai JCK. High-affinity transport of glutathione is part of a multicomponent system essential for mitochondrial function. Proceedings of the National Academy of Sciences of the United States of America. 2014;**87**:7185-7189. DOI: 10.1073/pnas.87.18.7185

[44] Wilkins HM, Kirchhof D, Manning E, Joseph JW, Linseman DA. Mitochondrial glutathione transport is a key determinant of neuronal susceptibility to oxidative and nitrosative stress. Biological Chemistry. 2013;**288**(7):5091-5101. DOI: 10.1074/jbc.M112.405738

[45] Stefanatos R, Sanz A. The role of mitochondrial ROS in the aging brain. FEBS Letters. 2018;**592**:743-758. DOI: 10.1002/1873-3468.12902

[46] Armstrong JS, Jones DP. Glutathione depletion enforces the mitochondrial permeability transition and causes cell death in Bcl-2 overexpressing HL60 cells. The FASEB Journal. 2002;**9**:1263-1265

[47] Packer L, Weber SU, Rimbach G. Molecular mechanisms of protective effects of molecular aspects of alpha-tocotrienol antioxidant action and cell signalling. The Journal of Nutrition.

2001;**131**:369S-373S. DOI: 10.1093/jn/131.2.369S

[48] Packer L, Weber SU, Rimbach G. Molecular aspects of alpha-tocotrienol antioxidant action and cell signalling. The Journal of Nutrition. 2001;**131**(2):369-373

[49] Abe K, Saito H. Characterization of t-butyl hydroperoxide toxicity in cultured rat cortical neurones and astrocytes. Pharmacology & Toxicology. 1998;**83**:40-46

[50] Mischel RE, Kim YS, Sheldon RA, Ferriero DM. Hydrogen peroxide is selectively toxic to immature murine neurons in vitro. Neuroscience Letters. 1997;**231**:17-20

[51] Whittemore ER. Peroxide-induced cell death in primary neuronal culture. Neuroscience. 1995;**67**(4):921-932

[52] Dringen R, Kussmaul L, Gutterer JM, Hirrlinger J, Hamprecht B. The glutathione system of peroxide detoxification is less efficient in neurons than in astroglial cells. Journal of Neurochemistry. 1999;**72**(6):2523-2530. DOI: 10.1046/j.1471-4159.1999.0722523.x

[53] Desagher S, Glowinski J, Premont J. Astrocytes protect neurons from hydrogen peroxide toxicity. The Journal of Neuroscience. 1996;**16**(8):2553-2562. DOI: 10.1523/jneurosci.16-08-02553.1996

[54] Nicole A, Ceballos-Picot I. Direct evidence for glutathione as mediator of apoptosis in neuronal cells. Biomedicine & Pharmacotherapy. 1998;**52**:349-355

[55] Franco R, Cidlowski JA. Apoptosis and glutathione: Beyond an antioxidant. Cell Death and Differentiation. 2009;**16**:1303-1314. DOI: 10.1038/cdd.2009.107

[56] Adriana Galvan TW. Pathophysiology of Parkinsonism. Clinical Neurophysiology. 2008;**119**(7):1459-1474. DOI: 10.1038/jid.2014.371

[57] Rodgers AD, Curzon G. Melanin formation by human brain. Journal of Neurochemistry. 1975;**14**(1965):1123-1129

[58] Hastings TG. Enzymatic oxidation of dopamine: The role of prostaglandin H synthase. Journal of Neurochemistry. 1995;**64**:919-924

[59] Kastner A, Hirsch EC, Agid Y. Is the vulnerability of neurons in the Substantia Nigra of patients with Parkinson's disease related to their neuromelanin content? Journal of Neurochemistry. 1992;**59**:1080-1089

[60] Dexter DT et al. Indices of & dative stress and mitochondrial function in individuals with incidental k w y body disease. Annals of Neurology. 1994;**35**:38-44

[61] Sian J, Dexter DT, Lees AJ, Daniel S, Jenner P, Marsden CD. Glutahone-related enzymes in brain in Parkinson' s disease. Annals of Neurology. 1994;**36**:356-361

[62] Perry TL, Yong VW. Idiopathic Parkinson's disease, progressive supranuclear palsy and glutathione metabolism in the substantia nigra of patients. Neuroscience Letters. 1986;**67**:269-274

[63] Shen X, Dryhurst G. Further insights into the influence of L-cysteine on the oxidation chemistry of dopamine: Reaction pathways of potential relevance to Parkinson's disease. Chemical Research in Toxicology. 1996;**9**:751-763

[64] Li H, Dryhurst G. Irreversible inhibition of mitochondrial complex I by benzothiazine-3-carboxylic acid (DHBT-1): A putative nigral endotoxin of relevance to Parkinson's disease. Journal of Neurochemistry. 1997;**69**:1530-1541

[65] Andersen JK. What causes the build-up of ubiquitin-containing inclusions in Parkinson's disease? Mechanisms of Ageing and Development. 2000;**118**:15-22

[66] Hershko A, Ciechanover A. The ubiquitin system. Annual Review of Biochemistry. 1998;**67**:425-479

[67] Jha N, Kumar MJ, Boonplueang R, Andersen JK. Glutathione decreases in dopaminergic PC12 cells interfere with the ubiquitin protein degradation pathway: Relevance for Parkinson's disease? Journal of Neurochemistry. 2002;**80**:555-561

[68] Bharath S, Hsu M, Kaur D, Rajagopalan S, Andersen JK. Glutathione, iron and Parkinson's disease. Biochemical Pharmacology. 2002;**64**:1037-1048

[69] Albers DS, Beal MF. Mitochondrial dysfunction and oxidative stress in aging and neurodegenerative disease. Journal of Neural Transmission. Supplementum. 2000;**59**:133-154

[70] Papers JBC et al. Glutathione depletion in PC12 results in selective inhibition of mitochondrial complex I activity. The Journal of Biological Chemistry. 2000;**275**(34):26096-26101. DOI: 10.1074/jbc.M000120200

[71] Davey GP, Peuchen S, John B, Davey GP, Peuchen S, Clark JB. Energy thresholds in brain mitochondria. Potential involvement in neurodegeneration. The Journal of Biological Chemistry. 1998;**273**:12753-12757. DOI: 10.1074/jbc.273.21.12753

[72] Haas RH et al. Low platelet mitochondrial complex I and complex II/III activity in early untreated Parhnson's disease. Annals of Neurology. 1995;**37**:714-722

[73] Lenaz G et al. Mitochondrial complex I defects in aging. Molecular and Cellular Biochemistry. 1997;**174**:329-333

[74] Mlay AI. Superoxide accelerates DNA damage by elevating free-iron levels. Proceedings of the National Academy of Sciences of the United States of America. 1996;**93**:13635-13640

[75] Ziegler DM. Role of reversible oxidation-reduction of enzyme thiols-disulfides in metabolic regulation. Annual Review of Biochemistry. 1985;**54**:305-329

[76] Fornstedt B, Bergh I, Rosengren E, Carlsson A. An improved HPLC-electrochemical detection method for measuring brain levels of 5-S-cysteinyldopamine, acid. Journal of Neurochemistry. 1990;**54**:578-586

[77] Chen JJ, Yu BPAL. Alterations in mitochondrial membrane fluidity by lipid peroxidation products. Free Radical Biology & Medicine. 1994;**17**(5):411-418

[78] Yoritaka MY, Hattori N, Uchida K, Tanaka M, Stadtman ER. protein Immunohistochemical detection of 4-hydroxynonenal adducts in Parkinson disease. Proceedings of the National Academy of Sciences of the United States of America. 1996;**93**:2696-2701

[79] Haass C, Selkoe DJ. Soluble protein oligomers in neurodegeneration: lessons from the Alzheimer's amyloid β -peptide. Molecular and Cellular Biology. 2007;**8**:101-112. DOI: 10.1038/nrm2101

[80] Cheignon C, Tomas M, Faller P, Hureau C, Collin F. Oxidative stress and the amyloid beta peptide in Alzheimer's disease. Redox Biology. 2018;**14**:450-464

[81] Hodgson N, Trivedi M, Muratore C, Li S, Deth R. Soluble oligomers of amyloid- NL Cause changes in redox state, DNA methylation, and gene transcription by inhibiting EAAT3

mediated cysteine uptake. Journal of Alzheimer's Disease. 2013;**36**:197-209. DOI: 10.3233/JAD-130101

[82] Duerson K et al. Detergent-insoluble EAAC1/EAAT3 aberrantly accumulates in hippocampal neurons of Alzheimer's disease patients. Brain Pathology. 2009;**19**:267-278. DOI: 10.1111/j.1750-3639.2008.00186.x

[83] Ansari SW, Scheff MA. Oxidative stress in the progression of Alzheimer disease in the frontal cortex. Journal of Neuropathology and Experimental Neurology. 2010;**69**(2):155-167. DOI: 10.1097/NEN.0b013e3181cb5af4.Oxidative

[84] Mandal PK, Tripathi M, Sugunan S. Brain oxidative stress: Detection and mapping of anti-oxidant marker 'Glutathione' in different brain regions of healthy male/female, MCI and Alzheimer patients using non-invasive magnetic resonance spectroscopy. Biochemical and Biophysical Research Communications. 2012;**417**:43-48. DOI: 10.1016/j.bbrc.2011.11.047

[85] Liu H, Harrell LE, Shenvi S, Hagen T, Liu R. Gender differences in glutathione metabolism in Alzheimer's disease. Neuroscience Research. 2005;**867**:861-867. DOI: 10.1002/jnr.20424

[86] Sultana DA, Piroddi R, Galli M, Butterfield F. Protein levels and activity of some antioxidant enzymes in hippocampus of subjects with amnestic mild cognitive impairment. Neurochemical Research. 2008;**33**: 2540-2546. DOI: 10.1007/s11064-008-9593-0

[87] Spalletta P, Bernardini G, Bellincampi S, Federici L, Trequattrini G, Ciappi A, et al. Glutathione S-transferase P1 and T1 gene polymorphisms predict longitudinal course and age at onset of Alzheimer disease. The American Journal of Geriatric Psychiatry. 2007;**15**:879-887. DOI: 10.1097/JGP.0b013e3180547076

[88] Casado R, Lopez-Fernandez AE, Casado MC, de la Torre M. Lipid peroxidation and antioxidant enzyme activities in vascular and Alzheimer dementias. Neurochemical Research. 2008;**33**:450-458. DOI: 10.1007/s11064-007-9453-3

[89] Butterfield DA, Castegna A, Lauderback CM, Drake J. Evidence that amyloid beta-peptide-induced lipid peroxidation and its sequelae in Alzheimer's disease brain contribute to neuronal death. Neurobiology of Aging. 2002;**23**:655-664

[90] Sultana R, Perluigi M. Protein oxidation and lipid peroxidation in brain of subjects neurodegeneration from redox proteomics. Antioxidants & Redox Signaling. 2006;**8**:2021-2037

[91] Islam T, Lynn BC, Markesbery WR, Lovell MA. Increased levels of 4-hydroxynonenal and acrolein, neurotoxic markers of lipid peroxidation, in the brain in mild cognitive impairment and early Alzheimer' s disease. Neurobiology of Aging. 2006;**27**:1094-1099. DOI: 10.1016/j.neurobiolaging.2005.06.004

Chapter 5

Periodontal Health and Disease in Glutathione Peroxidase

Figen Öngöz Dede

Abstract

Periodontal diseases are chronic, multifactorial inflammatory diseases that affect more than 10% of the world population. There are two general forms of periodontal diseases including gingivitis (reversible inflammation and confined with gingiva form) and periodontitis (irreversible, destruction form). Several studies have reported that periodontal disease was associated with a decreased antioxidant capacity and elevated oxidative damage within the oral cavity. Glutathione peroxidase (GSH-Px) is an important enzymatic antioxidant that protects periodontal tissues against oxidative stress. Hitherto, there is contradictory evidence concerning the relationship between the levels of GSH-Px and the periodontal status. Various studies have demonstrated that GSH-Px levels in different biological fluids increased, decreased, or are unaltered in individuals with periodontal disease. This discrepancy might be explained either by different determination protocols/assays applied among the studies or various dynamic processes of the periodontal disease progression. In this section, GSH-Px levels are summarized in the periodontal health and disease including the presence and absence of systemic disease, medication, wound healing, and smoking.

Keywords: glutathione peroxidase, periodontitis, gingivitis, gingival crevicular fluid, salivary

1. Introduction

The periodontium is a private connective tissue consisting of a gingiva, cementum, periodontal ligament, and alveolar bone supporting the tooth in the socket [1]. Periodontal disease is a widespread, chronic multimicrobial immuno-inflammatory illness which began with the complex coaction between the host's immunoinflammatory responses and pathogenic bacteria in the dental tissue [2]. There are two general forms of periodontal diseases including gingivitis (confined with gingiva form) and periodontitis. Gingivitis is a localized inflammation of the gingiva, which is began by pathogens in the microbial dental plaque on the tooth and gingiva [3]. Gingivitis causes reversible inflammation in the periodontal tissues [3]. Periodontitis, the destructive form of periodontal disease, leads to the destruction of the gingiva, alveolar bone, and periodontal ligament and is responsible for causing tooth mobility and early tooth loss [3, 4]. Periodontitis leads to irreversible local periodontal tissue destruction [5]. Periodontal diseases are the most common chronic diseases impacting 10–15% of population worldwide [6, 7].

Microbial dental plaque, mostly gram-negative anaerobic or facultative pathogens inside the subgingival biofilm, is the principal etiological factor in periodontal diseases [8]. Robust evidence in the etiology of periodontal diseases has been shown

IntechOpen

responsible for periodontopathogens including *Aggregatibacter actinomycetemcomitans* (Aa), *Porphyromonas gingivalis* (PG), *Tannerella forsythia* (TF), and *Treponema denticola* (TD) [9]. It is stated that "red complex" pathogens (PG, TD, and TF) are frequent in individuals with periodontitis [10]. The plurality of periodontal tissue devastation is brought about by an unsuitable host response to those pathogens and their products (lipopolysaccharides and proteases) [11]. The coaction between pathogenic bacteria and the host's immune response is participated by chemokines, the produce of pro-inflammatory cytokines, and an exaggerated immune response, entailing an increase in the number and activity of polymorphonuclear leukocytes (PMNs) [12]. PMNs are the main mediators of host response averse to the bacteria [13].

PMNs create the first advocacy of cellular host defenses averse to pathogenic microorganisms in the gingival sulcus [14]. PMNs defend the host against bacteria in two pathways, including oxygen-dependent and non-oxygen-dependent mechanisms [15]. The oxygen-dependent pathway contains the production of reactive oxygen species (ROS), which causes the destruction of periodontal tissues [16]. Although the main reason for the production of ROS by PMNs is the killing of bacteria, excessive production of ROS in the extracellular space causes the destruction of tissues [8, 14, 16]. The overproduction of ROS leads to tissue damage through different mechanisms including lipid peroxidation, DNA and protein damage, and the stimulation of pro-inflammatory cytokine [1, 16]. Several studies have shown the relationship between ROS and periodontal disease [5, 16–18]. Oxidative stress (OS), an imbalance between the pro-oxidant and antioxidant system, is involved in the bone resorptive process during periodontal disease [19]. Various studies have shown that OS is involved in the pathophysiological mechanisms of periodontitis [1, 17, 18, 20, 21]. Recently Sreeram et al. have described it as follows: "Periodontitis is an inflammatory condition leading to increased OS" [22].

Antioxidant defense mechanisms (nonenzymatic and enzymatic antioxidants) eliminate ROS and inhibit their detrimental consequences on the host [23]. Antioxidant enzymes protect tissues against the destructive effects of ROS created by different metabolic processes, modulating the dimension of inflammatory response [18, 24]. The defense mechanism averse to ROS involves three antioxidant pathways including intracellular, extracellular, and membrane antioxidants [25]. The main system is intracellular ROS cleaning enzymes: superoxide dismutase (SOD), catalase (CAT), and glutathione peroxidase (GSH-Px) [25]. GSH-Px, as a selenium-containing peroxidase, is a major group of enzymes that eliminate hydrogen peroxide created by SOD in the cytosol and mitochondria by oxidizing reduced glutathione to its oxidized form [22, 26]. GSH-Px is one of the enzymes that has a significant role in host defense averse to oxidative stress in cytosol [1, 18]. GSH-Px1 inhibits cytotoxic peroxide-induced oxidative damage, protein degeneration, and lipid peroxidation [27].

Traditional diagnosis of periodontal disease is based on clinical (gingival index (GI), bleeding on probing (BOP), clinical attachment level (CAL), probing pocket depth (PPD)) and radiographic parameters [22]. Traditional clinical measurements that are used for periodontal diagnosis are frequently of only restricted usefulness inasmuch as they are indicators of previous periodontal disease rather than present disease activity [28, 29]. Knowing the disease activity will enable early detection of the disease [21]. Moreover, the levels of oxidative stress parameters in saliva and gingival crevicular fluid (GCF) may show the activity and severity of periodontal disease [29].

Saliva is used as an easily collected diagnostic fluid that makes it possible to determine the levels of biomarkers in the evaluation of the disease condition [30]. By the way, GCF is a biological fluid in the gingival sulcus that derives from blood plasma and consists of metabolic elements of pathogens and host cells, which are explained as transudates or exudates [31]. Thereby, disease diagnosis via analysis of saliva and GCF is suitable for individuals [21]. Since the half-life of ROS is very short, they cannot be determined easily. Thus, ROS-induced demolition products and the activity of enzymatic and nonenzymatic antioxidants are optimal candidates to assess the consequences of OS-connected events in the pathological process of chronic periodontitis [21]. On the other hand, antioxidants (enzymatic and nonenzymatic antioxidants) in the saliva preserve the unity of oral tissues by neutralizing ROS [14].

In this section, GSH-Px levels are summarized in the periodontal health and disease including the presence and absence of systemic disease, medication, wound healing, smoking.

2. Glutathione peroxidase and periodontal disease

Wei et al. [32] examined the role of glutathione peroxidase in the pathogenesis of periodontal diseases. They reported higher total amount of GSH-Px in GCF samples from patients with gingivitis and periodontitis compared to healthy subjects [32]. Moreover, they determined that the total amount of GSH-Px was significantly higher in periodontitis patients than in gingivitis patients [32]. Also, there is a positive and significant correlation between the total amount of glutathione peroxidase and interleukin (IL)-1β and plaque index (PI) in GCF of the individuals with periodontal disease [32]. Besides, Panjamurthy et al. [33] assessed the levels of GSH-Px in patients with chronic periodontitis (CP) and determined that GSH-Px activities in the plasma, erythrocyte lysate, and gingival tissues were significantly increased in patients with periodontitis compared to healthy subjects. In addition, Borges et al. [34] analyzed the GSH-Px activities in the gingival tissue of individuals with CP. They determined a significant increase in GSH-Px activities in the individuals with CP when compared to the control group [34]. They noticed that an increase in GSH-Px may indicate possible antioxidant suppression in the destroyed ROS products in the gingival tissue [34]. Moreover, Arunachalam et al. [35] stated that the GSH-Px levels in the plasma of patients with aggressive periodontitis increased compared with the healthy individuals. Conversely, Sreeram et al. [22] and Aziz et al. [36] reported that serum GSH-Px activity in individuals with CP decreased when compared with the control groups.

Tsai et al. [37] aimed to determine the GSH-Px levels in saliva before and after periodontal treatment in patients with CP. They did not find a significant difference in the activities of GSH-Px in saliva between periodontally diseased and healthy subjects and even between prior to and after treatment in periodontitis patients [37]. On the contrary, Çanakçı et al. [20] and Miricescu et al. [38] found that the GSH-Px activities in saliva of patients with periodontitis were significantly lower than the controls. In accordance with Tsai et al. [37], Çanakcı et al. [20] suggested that there were no significant correlations between salivary GSH-Px capacities and periodontal status. Contrary to Tsai et al. [37], Novaković et al. [14, 39] evaluated the GSH-Px activity in saliva of the CP patients before and after nonsurgical treatment and concluded that there was a significant increase in these levels after therapy. Novaković et al. [14] argued that the increase in GSH-Px activity in saliva can be explained by the reduction in periodontal tissue inflammation after nonsurgical therapy. Novaković et al. [39] indicated that salivary GSH-Px could be used as a reliable biomarker in evaluating periodontal status and therapy outcome. A recent meta-analysis declared that there are no significant differences in the salivary GSH-Px levels between the patients with CP and periodontally healthy individuals [21]. These meta-analysis results coincide with other studies that have determined an increase or decrease in salivary GSH-Px levels [21]. The authors claimed that this disagreement might be connected to the various dynamic processes of the periodontitis progression [21].

Almerich-Silla et al. [17] showed the association between GSH-Px levels and the presence of different periodontal pathogens (PG, Aa, TD, and TF). They reported that GSH-Px levels were elevated in the existence of all bacteria types, except PG genotypes III and IV, and also the presence of different types of bacteria has a positive relationship with GSH-Px [17]. The authors advised that determination of GSH-Px levels and periodontal bacteria can be an important tool to control the progression of periodontal disease [17].

Until today, there is contradictory evidence concerning the relationship between the levels of GSH-Px and the periodontal status. Various studies have demonstrated that GSH-Px levels in different biological fluids increased, decreased, or are unaltered in individuals with periodontal disease. This discrepancy might be explained by different determination protocols/assays applied among the studies. On the other hand, a more persuasive change in GSH-Px activity in GCF than in saliva is observed [40]. GCF is more specific for periodontal inflammation than saliva, and also, saliva and serum GSH-Px levels may be affected by systemic conditions.

Replace the entirety of this text with the main body of your chapter. The body is where the author explains experiments, presents and interprets data of one's research. Authors are free to decide how the main body will be structured. However, you are required to have at least one heading. Please ensure that either British or American English is used consistently in your chapter.

2.1 Plasma glutathione peroxidase and periodontal disease

Patel et al. [8] examined the levels of plasma glutathione peroxidase (eGPx) in GCF before and 6–8 weeks after periodontal therapy in patients with periodontal disease. They ascertained that eGPx levels in GCF were significantly elevated progressively from health to gingivitis and periodontitis [8]. The study suggested that increased eGPx level in GCF from inflamed gingiva may indicate the increased ROS generation at the diseased site [8]. Also, the authors determined that the mean concentration of eGPx in GCF in CP group showed a significant reduction after the treatment and thus stated that increased eGPx concentration is associated with the severity of periodontal disease [8]. Similar to the previous study, Patel et al. [41] determined an increase of the eGPx concentrations in GCF and serum progressively from health to gingivitis and periodontitis groups and a decrease of these levels after nonsurgical periodontal therapy. Thus, the authors declared that the increase in GCF and serum eGPx can be considered as a marker of oxidative stress caused by periodontal infection [41]. Moreover, they noted that the significant increase in serum eGPx concentration in the periodontal disease can be possibly because of the overflow from the diseased periodontal tissues or increased production of eGPx by kidney proximal tubules in response to systemic oxidative stress caused by periodontal disease [41].

2.2 Glutathione peroxidase and periodontal wound healing

Sakallıoğlu et al. [25] investigated GSH-Px profiles in the 30-day recovery period (at days 3, 12, 21, and 30) in an acute incisional wound model created with mucoperiosteal periodontal flaps in dogs. They determined that GSH-Px levels increased significantly on the 3rd day of recovery period and then decreased insignificantly on the 12th day and increased insignificantly on the 21st day [25]. Later, GSH-Px levels decreased significantly on the 30th day compared to the 21st day of the recovery period, and these levels are lower than the baseline [25]. It is suggested that GSH-Px plays a significant role in the eradication of ROS in the recovery period of periodontal repair [25]. Moreover they argued that GSH-Px can neutralize to the noxious effects of OH in a normal periodontal mucoperiosteal or gingival wound healing [25].

2.3 Glutathione peroxidase, smoking, and periodontal disease

The etiology of periodontal disease is multifactorial, and periodontal pathogenesis processes are replaced by environmental and acquired risk factors such as smoking [42]. Tobacco smoking is one of the principal modifiable risk factor associated with chronic destructive periodontal disease [36]. It has been reported that the prevalence of periodontitis was three to six times higher in smokers than nonsmokers [16]. Possible negative effects of smoking on periodontal tissues may include altered neutrophil function, decreased IgG production, vascular alterations, increased prevalence of perio-pathogens, altered fibroblast attachment and functions, decreased lymphocyte proliferation, difficulty in eliminating pathogens by mechanical therapy, and negative local effects on cytokinesis and growth factor production [36]. Smoking influences oxidative stress in the body by promoting oxidative burst in neutrophils and causes an imbalance between antioxidants and ROS [43].

Guentsch et al. [16] evaluated both GSH-Px activities in saliva and serum in patients with periodontitis and the effects of periodontal treatment and smoking on these parameters. They reported an elevated GSH-Px activity in saliva in both the nonsmoking and smoking periodontitis groups compared to the periodontally healthy control groups and that these levels, which increased in both periodontitis groups, decreased after treatment [16]. However, the authors did not find a significant difference in serum GSH-Px values of both smokers and nonsmoker individuals with periodontitis and those who are periodontally healthy [16]. It is suggested that elevated GSH-Px levels in the saliva of periodontitis patients indicate to adversely affect antioxidant mechanisms leading to tissue damage of the continuous ROS production in periodontal inflammation [16]. Also, it is shown that smoking increased the GSH-Px levels in patients with periodontitis [16]. On the contrary, Aziz et al. [36] argued that smokers with CP have shown decreased GSH-Px activity in serum when compared to nonsmoker controls.

Hendek et al. [18] examined the effects of initial periodontal therapy on GSH-Px levels in serum, saliva, and GCF samples in smokers and nonsmokers with CP. They found that there was no significant difference among all groups for GSH-Px enzyme activity in serum, while GSH-Px enzyme activity in saliva and GCF was higher in smokers and nonsmokers with CP than periodontally healthy nonsmokers but statistically insignificant in GCF [18]. In addition, authors declared that there was no significant difference in the GSH-Px enzyme activity in GCF, serum, and saliva after periodontal therapy in both periodontitis groups [18]. Their data speculated that elevated GSH-Px activity in periodontitis patients may be a result of tissue repair and adaptive mechanisms against inflamed periodontal tissues in response to oxidative stress [18]. Conversely, Naresh et al. [43] found that the levels of GSH-Px in the saliva of smokers and nonsmokers with CP were decreased when compared with the healthy group and mean GSH-Px levels were lowest in smokers with CP. They stated that exposure to smoking may reduce salivary GSH-Px levels [43].

Toguç et al. [44] investigated the impact of smoking status on the GSH-Px levels in the gingival tissue and blood in subjects with CP. When blood GSH-Px levels are evaluated, the lowest values were observed in the smoker patients with CP compared to nonsmoker patients with CP and in the nonsmoker control group compared to nonsmokers and former smokers with CP [44]. Besides, elevated GSH-Px levels in gingival tissue have been determined in the control group when compared with all CP groups [44]. When gingival tissue GSH-Px levels are evaluated among all CP groups, the lowest values were found surprisingly in nonsmokers [44]. Moreover, they found that there were strong negative correlations between gingival tissue GSH-Px levels and smoking duration and yearly cigarette consumption [44]. Thus, they stated that the reduced local GSH-Px levels in the periodontitis patients may increase with smoking, and the reason for this increase may be the result of a protective and adaptive mechanism developing in the tissue [44].

2.4 Glutathione peroxidase, systemic diseases, and periodontal disease

Periodontal disease has been associated with several systemic illnesses, including atherosclerosis, cardiovascular disease, rheumatoid arthritis, diabetes mellitus, adverse pregnancy outcomes, and Alzheimer's disease [12].

2.4.1 Diabetes mellitus

Diabetes mellitus (DM) is a major risk factor for periodontal diseases, and periodontitis is noted as the sixth complication of DM. It has been shown with increasing evidence that the prevalence, progression, and severity of periodontitis increase in individuals with diabetes, especially uncontrolled, compared to individuals with no diabetes [45, 46]. There is a bilateral relationship between periodontal disease and DM. Various mechanisms have been suggested to clarify this relationship including the formation of advanced glycation end products (AGEs), changes in collagen metabolism and immune function, and recently an increased oxidative stress [47].

Arana et al. [48] evaluated the levels of GSH-Px in the saliva of patients with diabetes mellitus type 2 (DM2) and healthy nondiabetic patients in the presence of periodontal disease. They determined that the salivary GSH-Px levels in the diabetic group with good metabolic control was significantly higher than the control group and the diabetic group with poor metabolic control, and also patients with poor metabolic control in comparison with the control group and well-controlled diabetic groups have worst periodontal health and lowest saliva GPx levels [48]. Authors suggested that poor metabolic control in DM2 patients is associated with lower levels of salivary GSH-Px and worse periodontal health [48]. On the other hand, Duarte et al. [47] evaluated the gene expression of GSH-Px1 in the gingival tissue of poorly and well-controlled type 2 diabetic subjects with CP. They found that the periodontitis groups presented higher expression of GSH-Px1 than the periodontally healthy control [47]. It is advocated that GSH-Px1 was enhanced by periodontitis, independently of the diabetic status of the patients [47].

2.4.2 Cardiovascular disease

A recent review has shown a positive relationship between periodontitis and cardiovascular diseases [49]. It is determined that periodontal inflammation increases the development and progression of atheroma plaques via systemic bacteremia and lesion from the interaction of the intima with perio-pathogens entering the circulation [49]. Therefore, it is noted that the presence of periodontitis may be a risk factor for cardiovascular diseases [50]. Moreover, oxidative stress plays an important role in the pathogenesis of both periodontal disease and cardiovascular diseases [51].

Punj et al. [1] investigated the levels of glutathione peroxidase in serum and saliva of CP patients with and without ischemic heart disease (IHD). They stated that salivary GSH-Px levels were increased in the IHD + CP, IHD + H, and CP groups when compared with the healthy controls, whereas the serum GSH-Px levels were increased in the healthy group when compared with IHD + CP, IHD + H, and CP groups [1]. Authors indicated that this situation could probably be a result of a curative increase of GSH-Px to the oxidant stress in diseased states [1]. They emphasized that increased oxidative stress in the presence of chronic periodontitis may cause endothelial dysfunction of the blood vasculature, predisposing to atherosclerotic plaque formation and increasing predisposition to ischemic heart disease [1]. Köse et al. [52] examined the influences of periodontitis on levels of cardiac oxidative stress. Authors found that GSH-Px levels in the heart ventricular tissue of the rats with experimental periodontitis were higher than that of control group but statistically insignificant [52]. They argued that this increase could be associated with adaptive response [52]. Moreover, they speculated that oxidative stress in the cardiac tissue may be the result of an increase in the amount of ROS rather than a decrease in antioxidant levels [52].

2.4.3 Pregnancy

Various studies have proven a possible bidirectional association with periodontal disease and pregnancy [53]. It is supported that periodontal diseases are related with adverse pregnancy effects [54]. One of the possible mechanisms underlying this interaction stated that there may be oxidative stress-related inflammation pathways in case of pregnancy and periodontal disease [5, 27]. Oxidative stress is a principal supporting factor in the pathogenesis of preeclampsia and periodontal disease [27]. Çanakçı et al. [40] evaluated the GSH-Px levels in serum, saliva, and GCF in preeclamptic and normotensive pregnant women with and without periodontal disease. They determined that the GSH-Px activities in the serum and GCF of the periodontally healthy normotensive women were higher than that of preeclamptic and normotensive women with periodontal disease and periodontally healthy preeclamptic women [40]. There was no significant differences in saliva GSH-Px activities among all groups [40]. They declared that systemic and local GSH-Px activities reduced with the effect of periodontal disease in addition to the impact of preeclampsia [40]. Similarly Shetty et al. [27] observed that the GSH-Px activity in serum and saliva elevated in normotensive pregnant women with healthy periodontium when compared with preeclampsia pregnant women with and without periodontitis, and also preeclamptic women with periodontitis group have the lowest values but statistically nonsignificant. They indicated that periodontal diseases which cause a reduction in antioxidant levels could be a likelihood risk factor for severity, progression, and even initiation of preeclampsia [27].

Gümüş et al. [5] examined the salivary GSH-Px levels of the pregnant and postpartum women and their link with clinical parameters of periodontal inflammation and disease severity. They assigned that the GSH-Px levels were increased in the postpartum group when compared with pregnant and nonpregnant groups and in the nonpregnant group when compared with pregnant group [5]. Furthermore, they found that salivary GSH-Px levels were positively correlated with PD and BOP and total bacterial numbers in the postpartum group and with PD, CAL, BOP, or PI in the nonpregnant women group [5]. Conversely authors did not find association between GPx levels and periodontal disease status in pregnant women [5]. It is determined that salivary GSH-Px levels, which were at low levels during pregnancy, increased in the postpartum [5]. They speculated that this may be due to a healing mechanism against the exposure of tissues to excessive ROS during pregnancy [5].

2.5 Glutathione peroxidase, medication, and periodontal disease

Drug-induced gingival enlargement is previously reported as side effect of immunosuppressive agents such as cyclosporine A (CsA) and tacrolimus, calcium channel blockers such as amlodipine and nifedipine, and anticonvulsant drugs such as phenytoin [55]. It has been stated that overgrowth develops due to the increase in collagen accumulation and decrease in collagenase enzyme activity after drug use [55]. Gingival and periodontal inflammation may increase, as excessive gingival enlargement will complicate oral hygiene practices [55]. Sobeniec et al. [56] evaluated the GSH-Px activity in serum and saliva in patients with periodontal disease treated due to epilepsy. They determined that serum and saliva GSH-Px activities decreased in these patients with excessive gingival enlargement when compared with the control group [56]. On the other hand, Sardarian et al. [26], an in vitro study, determined that the low concentration of CsA (0.1 mg/mL) had no effect on GSH-Px activity in the oral epithelium while the activity was significantly increased at higher concentrations (1 mg/mL). They argued that GSH-Px activity increased to eliminate increased ROS in the oral epithelium after treatment with CsA [26].

In an experimental study, rats were infected with multibacterial inoculum containing PG, TD, and TF, as an oral lavage every other week for 12 weeks [12]. Afterward, daily subcutaneous injections of enoxacin, bis-enoxacin, alendronate, or doxycycline were administered for 6 weeks after 6 weeks of multibacterial infection in rats [12]. Subsequently, they evaluated the levels of GSH-Px in the serum of the infected, treated, and sham-infected rats [12]. Consequently, it is determined that serum levels of GSH-Px increased in rats infected with periodontal bacteria when compared with sham-infected rats and reduced in treated rats compared to infected and untreated rats [12]. Authors stated that elevated GSH-Px activity protects the periodontal tissues averse to oxidative stress [12].

Host modulatory therapy (HMT) is a treatment method that aims to decrease tissue destruction and stabilize the periodontium by arranging the components of the host response [57]. HMTs may be categorized as anti-inflammatory drugs, bone-stimulating agents (bisphosphonates), and anti-proteinase agents, such as low-dose doxycycline (LDD) [58]. Caffeic acid phenethyl ester (CAPE) has antioxidant, antitumoral, anti-inflammatory, and immunomodulatory properties and inhibits ROS production during inflammatory processes [59]. Recently, it has been reported that CAPE can modulate the host response [60]. Yiğit et al. [19] evaluated the effects of LDD and CAPE on alveolar bone level and the plasma levels of GSH-Px activity in an experimental periodontitis rat model. They determined that GSH-Px levels in plasma increased in the CAPE + periodontitis group, but decreased in the periodontitis and periodontitis + LDD groups when compared to control group [19]. The authors stated that CAPE significantly increased GSH-Px levels and CAPE may have more antioxidant properties than LDD in periodontal inflammation [19].

A previous study showed the creation of fast reepithelization on the human gingival wounds of the topical application of 1% taurine (2-amino ethane sulfonic acid) [61]. Sree and Sethupathy [62] investigated the effect of taurine as an antioxidant in the management of patients with the chronic periodontitis. For this purpose, they evaluated GSH-Px levels in the plasma and gingival tissue before and after administration of taurine [62]. They reported that decreased GSH-Px levels in plasma and gingival tissue were determined after taurine administration [62]. It is suggested that taurine enhanced the antioxidant status of chronic periodontitis patients by affecting GSH-Px antioxidant levels [62].

While melatonin has a direct neutralizing effect against ROS, it has an indirect effect by increasing the effectiveness of GSH-Px [63]. Özdem et al. [64] investigated the GSH-Px levels in the heart tissues after melatonin application after induction of experimental periodontitis in the rats. They found that the GSH-Px levels in heart tissue were higher in the periodontitis + melatonin group compared to periodontitis + saline solution group and in the healthy + melatonin group compared to healthy + saline solution group, while there were no significant differences between healthy + saline solution and periodontitis + saline solution groups [64]. In line with these results, the authors claimed that application of melatonin caused an increase in GSH-Px levels in the heart tissue either due to its antioxidant properties or by increasing the synthesis of antioxidant enzymes [64]. Furthermore, Kırzıoğlu et al. [24] examined the effects of systemically administered rosuvastatin, which decreases the levels of ROS and increases antioxidant activity, on GSH-Px levels in the serum of the rats with experimental periodontitis. They reported there were no significant differences in the levels of GSH-Px among control, healthy + rosuvastatin, periodontitis, and periodontitis + rosuvastatin groups [24].

3. Conclusions

There is a growing evidence for the role of ROS in the pathogenesis of periodontal diseases. The short half-life of ROS limits its measurability in biological fluids in the periodontal disease. Therefore, it is stated that it is more reliable to measure the products of ROS-induced tissue damage and levels of antioxidants in the periodontal disease. One of the most frequently detected enzymatic antioxidants in periodontal disease is GSH-Px. Previous studies found that GSH-Px levels in different biological fluids increased, decreased, or are unchanged in individuals with periodontal disease compared to control groups. The reason for this contradiction might be linked to the difference in the analyses applied between studies and the presence of various dynamic processes in progression of periodontal disease. Nevertheless, the common result in the studies stated that GSH-Px protects periodontal tissues against oxidative stress and plays an important role in the progression of periodontal disease. Thus, it was emphasized that GSH-Px can be a reliable biomarker in biological fluids to evaluate periodontal status and results of periodontal treatment. However, further studies in long term using large population are needed in order to better understand how GSH-Px contributes to the development of periodontal diseases using knockout and knockdown techniques.

Conflict of interest

The authors declare no conflict of interest.

Author details

Figen Öngöz Dede
Faculty of Dentistry, Department of Periodontology, Ordu University, Ordu, Turkey

*Address all correspondence to: figen_ongoz@hotmail.com

References

[1] Punj A, Shenoy S, Kumari NS, Pampani P. Estimation of antioxidant levels in saliva and serum of chronic periodontitis patients with and without ischemic heart disease. International Journal of Dentistry. 2017;**2017**:1965697. DOI: 10.1155/2017/1965697

[2] Ebersole JL, Dawson D 3rd, Emecen-Huja P, Nagarajan R, Howard K, Grady ME, et al. The periodontal war: Microbes and immunity. Periodontology 2000. 2017;**75**(1):52-115. DOI: 10.1111/prd.12222

[3] Kinane DF, Stathopoulou PG, Papapanou PN. Periodontal diseases. Nature Reviews. Disease Primers. 2017;**22**(3):17038. DOI: 10.1038/nrdp.2017.38

[4] Loesche WJ, Grossman NS. Periodontal disease as a specific, albeit chronic, infection: Diagnosis and treatment. Clinical Microbiology Reviews. 2001;**14**(4):727-752. DOI: 10.1128/CMR.14.4.727-752.2001

[5] Gümüş P, Emingil G, Öztürk VÖ, Belibasakis GN, Bostanci N. Oxidative stress markers in saliva and periodontal disease status: Modulation during pregnancy and postpartum. BMC Infectious Diseases. 2015;**8**(15):261. DOI: 10.1186/s12879-015-1003-z

[6] Ridgeway EE. Periodontal disease: Diagnosis and management. Journal of the American Academy of Nurse Practitioners. 2000;**12**(3):79-84. DOI: 10.1111/j.1745-7599.2000.tb00171.x

[7] Baelum V, Lopez R. Periodontal epidemiology: Towards social science or molecular biology? Community Dentistry and Oral Epidemiology. 2004;**32**(4):239-249. DOI: 10.1111/j.1600-0528.2004.00159.x

[8] Patel SP, Pradeep AR, Chowdhry S. Crevicular fluid levels of plasma glutathione peroxidase (eGPx) in periodontal health and disease. Archives of Oral Biology. 2009;**54**(6):543-548. DOI: 10.1016/j.archoralbio.2009.02.002

[9] Genco RJ. Current view of risk factors for periodontal diseases. Journal of Periodontology. 1996;**67**:1041-1049. DOI: 10.1902/jop.1996.67.10.1041

[10] Wara-aswapati N, Pitiphat W, Chanchaimongkon L, Taweechaisupapong S, Boch JA, Ishikawa I. Red bacterial complex is associated with the severity of chronic periodontitis in a Thai population. Oral Diseases. 2009;**15**(5):354-359. DOI: 10.1111/j.1601-0825.2009.01562.x

[11] Lamster IB, Novak MJ. Host mediators in gingival crevicular fluid: Implications for the pathogenesis of periodontal disease. Critical Reviews in Oral Biology and Medicine. 1992;**3**:31-60. DOI: 10.1177/10454411920030010501

[12] Oktay S, Chukkapalli SS, Rivera-Kweh MF, Velsko IM, Holliday LS, Kesavalu L. Periodontitis in rats induces systemic oxidative stress that is controlled by bone-targeted antiresorptives. Journal of Periodontology. 2015;**86**(1):137-145. DOI: 10.1902/jop.2014.140302

[13] Miller DR, Lamster IB, Chasens AI. Role of the polymorphonuclear leukocyte in periodontal health and disease. Journal of Clinical Periodontology. 1984;**11**:1-15. DOI: 10.1111/j.1600-051x.1984.tb01303.x

[14] Novaković N, Cakić S, Todorović T, Raicević BA, Dozić I, Petrović V, et al. Antioxidative status of saliva before and after non-surgical periodontal treatment. Srpski Arhiv za Celokupno Lekarstvo. 2013;**141**(3-4):163-168. DOI: 10.2298/sarh1304163n

[15] Halliwell B. Oral inflammation and reactive species: A missed opportunity?

Oral Diseases. 2000;**6**(3):136-137. DOI: 10.1111/j.1601-0825.2000.tb00324.x

[16] Guentsch A, Preshaw PM, Bremer-Streck S, Klinger G, Glockmann E, Sigusch BW. Lipid peroxidation and antioxidant activity in saliva of periodontitis patients: Effect of smoking and periodontal treatment. Clinical Oral Investigations. 2008;**12**(4):345-352. DOI: 10.1007/s00784-008-0202-z

[17] Almerich-Silla JM, Montiel-Company JM, Pastor S, Serrano F, Puig-Silla M, Dasí F. Oxidative stress parameters in saliva and its association with periodontal disease and types of bacteria. Disease Markers. 2015;**2015**:653537. DOI: 10.1155/2015/653537

[18] Hendek MK, Erdemir EO, Kisa U, Ozcan G. Effect of initial periodontal therapy on oxidative stress markers in gingival crevicular fluid, saliva, and serum in smokers and non-smokers with chronic periodontitis. Journal of Periodontology. 2015;**86**(2):273-282. DOI: 10.1902/jop.2014.140338

[19] Yiğit U, Kırzıoğlu FY, Uğuz AC, Nazıroğlu M, Özmen Ö. Is caffeic acid phenethyl ester more protective than doxycycline in experimental periodontitis? Archives of Oral Biology. 2017;**81**:61-68. DOI: 10.1016/j.archoralbio.2017.04.017

[20] Çanakçı CF, Cicek Y, Yildirim A, Sezer U, Çanakçı V. Increased levels of 8-hydroxydeoxyguanosine and malondialdehyde and its relationship with antioxidant enzymes in saliva of periodontitis patients. The European Journal of Dentistry. 2009;**3**(2):100-106

[21] Chen M, Cai W, Zhao S, Shi L, Chen Y, Li X, et al. Oxidative stress-related biomarkers in saliva and gingival crevicular fluid associated with chronic periodontitis: A systematic review and meta-analysis. Journal of Clinical Periodontology. 2019;**46**(6):608-622. DOI: 10.1111/jcpe.13112

[22] Sreeram M, Suryakar AN, Dani NH. Is gamma-glutamyl transpeptidase a biomarker for oxidative stress in periodontitis? Journal of Indian Society of Periodontology. 2015;**19**(2):150-154. DOI: 10.4103/0972-124X.149032

[23] Sculley DV, Langley-Evans SC. Periodontal disease is associated with lower antioxidant capacity in whole saliva and evidence of increased protein oxidation. Clinical Science (London, England). 2003;**105**(2):167-172. DOI: 10.1042/CS20030031

[24] Kırzıoğlu FY, Tözüm Bulut M, Doğan B, Fentoğlu Ö, Özmen Ö, Çarsancaklı SA, et al. Anti-inflammatory effect of rosuvastatin decreases alveolar bone loss in experimental periodontitis. Journal of Oral Science. 2017;**59**(2):247-255. DOI: 10.2334/josnusd.16-0398

[25] Sakallioğlu U, Aliyev E, Eren Z, Akşimşek G, Keskiner I, Yavuz U. Reactive oxygen species scavenging activity during periodontal mucoperiosteal healing: An experimental study in dogs. Archives of Oral Biology. 2005;**50**(12):1040-1046. DOI: 10.1016/j.archoralbio.2005.03.012

[26] Sardarian A, Andisheh Tadbir A, Zal F, Amini F, Jafarian A, Khademi F, et al. Altered oxidative status and integrin expression in cyclosporine A-treated oral epithelial cells. Toxicology Mechanisms and Methods. 2015;**25**(2):98-104. DOI: 10.3109/15376516.2014.990595

[27] Shetty MS, Ramesh A, Shetty PK, Agumbe P. Salivary and serum antioxidants in women with preeclampsia with or without periodontal disease. Journal of Obstetrics and Gynaecology of India. 2018;**68**(1):33-38. DOI: 10.1007/s13224-017-0993-4

[28] AlMoharib HS, AlMubarak A, AlRowis R, Geevarghese A,

Preethanath RS, Anil S. Oral fluid based biomarkers in periodontal disease: Part 1. Saliva. Journal of International Oral Health. 2014;**6**(4):95-103

[29] Baltacıoğlu E1, Yuva P, Aydın G, Alver A, Kahraman C, Karabulut E, Akalın FA. Lipid peroxidation levels and total oxidant/antioxidant status in serum and saliva from patients with chronic and aggressive periodontitis. Oxidative stress index: A new biomarker for periodontal disease? Journal of Periodontology. 2014;**85**(10):1432-1441. DOI: 10.1902/jop.2014.130654

[30] Trivedi S, Lal N, Mahdi AA, Singh B, Pandey S. Association of salivary lipid peroxidation levels, antioxidant enzymes, and chronic periodontitis. The International Journal of Periodontics & Restorative Dentistry. 2015;**35**(2):e14-e19. DOI: 10.11607/prd.2079

[31] Goodson JM. Gingival crevice fluid flow. Periodontology 2000. 2003;**31**:43-54. DOI: 10.1034/j.1600-0757.2003.03104.x

[32] Wei PF, Ho KY, Ho YP, Wu YM, Yang YH, Tsai CC. The investigation of glutathione peroxidase, lactoferrin, myeloperoxidase and interleukin-1beta in gingival crevicular fluid: Implications for oxidative stress in human periodontal diseases. Journal of Periodontal Research. 2004;**39**(5):287-293. DOI: 10.1111/j.1600-0765.2004.00744.x

[33] Panjamurthy K, Manoharan S, Ramachandran CR. Lipid peroxidation and antioxidant status in patients with periodontitis. Cellular & Molecular Biology Letters. 2005;**10**(2):255-264

[34] Borges I Jr, Moreira EA, Filho DW, de Oliveira TB, da Silva MB, Fröde TS. Proinflammatory and oxidative stress markers in patients with periodontal disease. Mediators of Inflammation. 2007;**2007**:45794. DOI: 10.1155/2007/45794

[35] Arunachalam R, Rajeev V, Kumaresan R, Kurra SB. Clinical and biochemical valuation of enzymatic and nonenzymatic stress markers following full-mouth disinfection in aggressive periodontitis. The Journal of Contemporary Dental Practice. 2019;**20**(8):952-956

[36] Aziz AS, Kalekar MG, Suryakar AN, Benjamin T, Prakashan MJ, Ahmed BM, et al. Assessment of some biochemical oxidative stress markers in male smokers with chronic periodontitis. Indian Journal of Clinical Biochemistry. 2013;**28**(4): 374-380. DOI: 10.1007/s12291-012-0283-y

[37] Tsai CC, Chen HS, Chen SL, Ho YP, Ho KY, Wu YM, et al. Lipid peroxidation: A possible role in the induction and progression of chronic periodontitis. Journal of Periodontal Research. 2005;**40**(5):378-384. DOI: 10.1111/j.1600-0765.2005.00818.x

[38] Miricescu D, Totan A, Calenic B, Mocanu B, Didilescu A, Mohora M, et al. Salivary biomarkers: Relationship between oxidative stress and alveolar bone loss in chronic periodontitis. Acta Odontologica Scandinavica. 2014;**72**(1):42-47. DOI: 10.3109/00016357.2013.795659

[39] Novaković N, Todorovic T, Rakic M, Milinkovic I, Dozic I, Jankovic S, et al. Salivary antioxidants as periodontal biomarkers in evaluation of tissue status and treatment outcome. Journal of Periodontal Research. 2014;**49**(1): 129-136. DOI: 10.1111/jre.12088

[40] Çanakçı V, Yildirim A, Çanakçı CF, Eltas A, Cicek Y, Çanakçı H. Total antioxidant capacity and antioxidant enzymes in serum, saliva, and gingival crevicular fluid of preeclamptic women with and without periodontal disease. Journal of Periodontology. 2007;**78**(8):1602-1611. DOI: 10.1902/jop.2007.060469

[41] Patel SP, Rao NS, Pradeep AR. Effect of nonsurgical periodontal

therapy on crevicular fluid and serum glutathione peroxidase levels. Disease Markers. 2012;**32**(1):1-7. DOI: 10.3233/DMA-2012-0855

[42] Johnson GK, Hill M. Cigarette smoking and the periodontal patient. Journal of Periodontology. 2004;**75**: 196-209. DOI: 10.1902/jop.2004.75.2.196

[43] Naresh CK, Rao SM, Shetty PR, Ranganath V, Patil AS, Anu AJ. Salivary antioxidant enzymes and lipid peroxidation product malondialdehyde and sialic acid levels among smokers and non-smokers with chronic periodontitis-A clinico-biochemical study. Journal of Family Medicine and Primary Care. 2019;**8**(9):2960-2964. DOI: 10.4103/jfmpc.jfmpc_438_19

[44] Tonguç MÖ, Öztürk O, Sütçü R, Ceyhan BM, Kılınç G, Sönmez Y, et al. The impact of smoking status on antioxidant enzyme activity and malondialdehyde levels in chronic periodontitis. Journal of Periodontology. 2011;**82**(9):1320-1328. DOI: 10.1902/jop.2011.100618

[45] Kinane D, Bouchard P. Group E of European workshop on periodontology periodontal diseases and health: Consensus report of the sixth European workshop on periodontology. Journal of Clinical Periodontology. 2008;**35**:333-337. DOI: 10.1111/j.1600-051X.2008.01278.x

[46] Awartani FA. Evaluation of the relationship between type 2 diabetes and periodontal disease. Saudi Medical Journal. 2009;**30**:902-906

[47] Duarte PM, Napimoga MH, Fagnani EC, Santos VR, Bastos MF, Ribeiro FV, et al. The expression of antioxidant enzymes in the gingivae of type 2 diabetics with chronic periodontitis. Archives of Oral Biology. 2012;**57**(2):161-168. DOI: 10.1016/j.archoralbio.2011.08.007

[48] Arana C, Moreno-Fernández AM, Gómez-Moreno G, Morales-Portillo C, Serrano-Olmedo I, de la Cuesta Mayor MC, et al. Increased salivary oxidative stress parameters in patients with type 2 diabetes: Relation with periodontal disease. Endocrinología, Diabetes y Nutrición. 2017;**64**(5):258-264. DOI: 10.1016/j.endinu.2017.03.005

[49] Carrizales-Sepúlveda EF, Ordaz-Farías A, Vera-Pineda R, Flores-Ramírez R. Periodontal disease, systemic inflammation and the risk of cardiovascular disease. Heart, Lung & Circulation. 2018;**27**(11):1327-1334. DOI: 10.1016/j.hlc.2018.05.102

[50] Lockhart PB, Bolger AF, Papapanou PN, Osinbowale O, Trevisan M, Levison ME, et al. American Heart Association rheumatic fever, endocarditis, and Kawasaki disease Committee of the Council on cardiovascular disease in the young, council on epidemiology and prevention, council on peripheral vascular disease, and council on clinical cardiology. Periodontal disease and atherosclerotic vascular disease: Does the evidence support an independent association? A scientific statement from the American Heart Association. Circulation. 2012;**125**(20):2520-2544. DOI: 10.1161/CIR.0b013e31825719f3

[51] Kurita-Ochiai T, Jia R, Cai Y, Yamaguchi Y, Yamamoto M. Periodontal disease-induced atherosclerosis and oxidative stress. Antioxidants (Basel). 2015;**4**(3):577-590. DOI: 10.3390/antiox4030577

[52] Köse O, Arabacı T, Yemenoglu H, Ozkanlar S, Kurt N, Gumussoy I, et al. Influence of experimental periodontitis on cardiac oxidative stress in rats: A biochemical and histomorphometric study. Journal of Periodontal Research. 2017;**52**(3):603-608. DOI: 10.1111/jre.12428

[53] Armitage GC. Bi-directional relationship between pregnancy and periodontal disease. Periodontology 2000. 2013;**61**:160-176. DOI: 10.1111/j.1600-0757.2011.00396.x

[54] Agueda A, Ramon JM, Manau C, Guerrero A, Echeverra JJ. Periodontal disease as a risk factor for adverse pregnancy outcomes: A prospective cohort study. Journal of Clinical Periodontology. 2008;**35**:16-22. DOI: 10.1111/j.1600-051X.2007.01166.x

[55] Brown RS, Arany PR. Mechanism of drug-induced gingival overgrowth revisited: A unifying hypothesis. Oral Diseases. 2015;**21**(1):e51-e61. DOI: 10.1111/odi.12264

[56] Sobaniec H, Sobaniec W, Sendrowski K, Sobaniec S, Pietruska M. Antioxidant activity of blood serum and saliva in patients with periodontal disease treated due to epilepsy. Advances in Medical Sciences. 2007;**52**(Suppl 1):204-206

[57] Preshaw PM. Host response modulation in periodontics. Periodontology 2000. 2008;**48**:92-110. DOI: 10.1111/j.1600-0757.2008.00252.x

[58] Yağan A, Kesim S, Liman N. Effect of low-dose doxycycline on serum oxidative status, gingival antioxidant levels, and alveolar bone loss in experimental periodontitis in rats. Journal of Periodontology. 2014;**85**(3):478-489. DOI: 10.1902/jop.2013.130138

[59] Akyol S, Ginis Z, Armutcu F, Ozturk G, Yigitoglu MR, Akyol O. The potential usage of caffeic acid phenethyl ester (CAPE) against chemotherapy-induced and radiotherapy-induced toxicity. Cell Biochemistry and Function. 2012;**30**(5):438-443. DOI: 10.1002/cbf.2817

[60] Choi EY, Choe SH, Hyeon JY, Choi JI, Choi IS, Kim SJ. Effect of caffeic acid phenethyl ester on Prevotella intermedia lipopolysaccharide-induced production of proinflammatory mediators in murine macrophages. Journal of Periodontal Research. 2015;**50**(6):737-747. DOI: 10.1111/jre.12260

[61] Gültekin SE, Sengüven B, Sofuoğlu A, Taner L, Koch M. Effect of the topical use of the antioxidant taurine on the two basement membrane proteins of regenerating oral gingival epithelium. Journal of Periodontology. 2012;**83**:127-134. DOI: 10.1902/jop.2011.100568

[62] Sree SL, Sethupathy S. Evaluation of the efficacy of taurine as an antioxidant in the management of patients with chronic periodontitis. Dental Research Journal (Isfahan). 2014;**11**(2):228-233

[63] Bonnefont-Rousselot D, Collin F. Melatonin: Action as antioxidant and potential applications in human disease and aging. Toxicology. 2010;**278**(1): 55-67. DOI: 10.1016/j.tox.2010.04.008

[64] Özdem M, Kırzıoğlu FY, Yılmaz HR, Vural H, Fentoğlu Ö, Uz E, et al. Antioxidant effects of melatonin in heart tissue after induction of experimental periodontitis in rats. Journal of Oral Science. 2017;**59**(1): 23-29. DOI: 10.2334/josnusd.16-0034

Section 2

Oxidative Stress in Health and Disease

Chapter 6

Effect of Oxidative Stress on Sperm Cells

Alejandro Córdova Izquierdo, Adrian Emmanuel Iglesias Reyes, Alda Roció Ortiz Muñiz, María de Lourdes Juárez Mosqueda, Jesús Alberto Guevara González, Juan Eulogio Guerra Liera, Gustavo Ruiz Lang, Rubén Huerta Crispín, Abel E. Villa Mancera, Pedro Sánchez Aparico and Raúl Sánchez Sánchez

Abstract

Free radicals are unstable molecules that have an unpaired electron in their last orbital, which makes them highly unstable agents. In medicine, it has been discovered that they play an important role in cell signaling and without them some cells such as leukocytes or sperm could not perform their biological functions. To protect itself from these oxidizing agents, the cell has a defense system based on antioxidants; however, when this balance is lost and oxidizing agents exceed the cellular antioxidant capacity, the cell enters oxidative stress, which affects cellular components such as proteins, nucleic acids, lipids, amino acids, and carbohydrates, among others. In the case of spermatozoa, due to their high metabolic rate, they produce large quantities of oxygen reactive species (ROS), decreasing sperm motility, alterations in cytoplasmic components, modifications in genetic material, or sperm death. In this chapter, a review is made of a brief history of how the toxicity of oxygen and free radicals was discovered, the oxidative stress in cells, and the effect of oxidative stress in the cytoplasmic sperm membrane, in the spermatic mitochondria, in the spermatic acrosome, in the sperm DNA, and in the fertility of the female and the male.

Keywords: spermatozoa, oxidative stress, free radicals, reproduction

1. Introduction

Semen freezing is one of the most important procedures in the development of biotechnologies for assisted reproduction. Among the advantages that we can find in artificial insemination is as follows: to keep the biological material viable for an indefinite time, the establishment of gene banks and the exchange of genetic material over very long distances economically rationalize the ejaculate; improve the use of wild boar elite, an adequate available germinal material of economic interest for man; and perform the collection of semen only in the most favorable reproductive seasons. However, the composition of the plasma membrane of the pig sperm, the large phospholipid layer (the comparison of bull sperm, which has a smaller layer of skin), is the cause of the sperm cell. Free radical changes that occur during freezing, the occasion when the effects of sperm freezing occur in the wild boar, affect the integrity of the plasma membrane, the acrosome, the nucleus, as well as the mitochondrial functions and motility of spermatozoa [1–4]. The purpose of this review is to publicize the main causes of ROS generation in sperm cells, as well as a brief explanation of how ROS is a part of sperm parts.

2. Background

Air is a vital element for any living being and is a mixture of gases based on nitrogen (78%), oxygen (21%), water vapor (variable between 0 and 7%), ozone, carbon dioxide (CO_2), hydrogen, and some noble gases such as krypton, neon, helium, and argon. Of these, oxygen (which appeared approximately 2500 million years ago) plays a vital role in the processes of aerobic life, being the second most abundant element in the atmosphere [5–7].

Antoine Lavoisier in the eighteenth century gives the name to "oxygen" which means "generator of acids," because despite having a therapeutic use, it was already known that it was a toxic substance, due to its great oxidizing power. In 1774, the toxic effects of the gas are demonstrated, and 6 years later (1780) experiments are made of the use of oxygen in newborns; in 1878, the toxic effect of oxygen in the brain is documented by Paul Bert, manifested by the presence of convulsive crises to more than three atmospheres, and in 1899, when trying to replicate the Bert effect, J. Lorrain Smith reports fatal pneumonia in rats exposed to 73% oxygen for 4 days. In 1940, it is reported that babies with periodic breathing pattern improved with the use of oxygen to 70%, beginning the routine use of oxygen in premature babies. Between 1951 and 1956, it is demonstrated that oxygen was safe when it occurred in concentrations lower than 40%. Harman in 1954 stated that the life expectancy increases decreasing the degree of oxidative phenomena. Thus, throughout history, it has been described that the higher the toxicity of O_2 is, the higher is the metabolic rate of the species considered [6, 8].

In veterinary and human medicine, more and more agents that cause diseases in the body have been discovered; some of them are derived from metabolic processes of oxygen, among which are the production of energy, detoxification of harmful compounds, and defense against pathogens, among which are free radicals (RL), which are highly reactive oxidation agents, which act as short-lived chemical intermediates on lipids, amino acids, carbohydrates, and nucleic acids [5, 7].

The RL can be divided into the following: (i) reactive oxygen species (ROS), which are highly reactive molecules that constantly attack organisms through oxidation-reduction reactions, among which are molecular oxygen (O_2), superoxide anion (O21) hydrogen peroxide (H_2O_2), hydroperoxyl (HO_2), and hydroxyl radical (OH); (ii) the transition metals, which have unpaired electrons and can exist as RL; and (iii) reactive nitrogen species (ERN), which are capable of generating oxidative damage and cell death, among which are nitric oxide (NO), peroxynitrite anion (ONOO−), and nitric dioxide (NO_2) [9–11].

The RL must be attenuated by different antioxidant defense systems, which involve enzymes and molecules. Antioxidants are divided into enzymatic, also called endogenous production, which are the first line of defense against the production of RL and are proteins with antioxidant capacity that are not consumed when reacting with the RL. Among the most important of this group are catalase, superoxide dismutase, and glutathione peroxidase. The nonenzymatic ones come mainly from the diet and are small liposoluble molecules, which, unlike the enzymatic, are consumed during their antioxidant action, so they must be replaced; among the most important in this group are vitamins E and C, beta-carotenes, retinol, uric acid, pyruvate, albumin, carnitine, taurine, hypotaurine, transferrin, ceruloplasmin, polyphenoids, flavonoids, and trace elements [12–16]. These antioxidant defense systems are linked in a cellular buffer system, where they add up and collaborate with each other, to deal with any oxidative aggression in cells, for example, nonenzymatic antioxidants can have synergistic effects in combination with enzymatic antioxidants, regenerating enzymatic antioxidants through the donation of hydrogen, neutralizing molecular oxygen, and catalyzing the synthesis or regeneration of nonenzymatic antioxidants [9].

When there is an imbalance and the amount of RL exceeds the balance between oxidant production and antioxidant capacity, a phenomenon known as oxidative stress (EO) is generated, which has negative consequences on multiple cellular processes [7, 14, 17, 18].

3. Effect of oxidative stress on cells

Due to aerobic conditions, cells maintain a high concentration of oxidant products in their metabolism, such as RL, which are generated as a result of cellular metabolism and in cellular physiological concentrations are related to cell signaling processes or to fulfill their functions biological, including leukocytes that are recruited to the sites of infection by chemotactic factors and are able to eliminate microorganisms through phagocytosis, exposing them to high concentrations of ROS (superoxide and hydrogen peroxide) and other microbicidal products contained in cell granules. However, when EO exists, ROS can mainly affect cellular components such as proteins, nucleic acids, sugars, and lipids [7, 9, 17].

Most of the main diseases that cause the death of animals and people or deteriorate their quality of life are caused by the RL. Each cell of the body suffers about 10,000 impacts of free radicals per day. For this reason, the EO has been the target of intense research in recent years, mainly in the implications on how mitochondria produce ROS, since they are of vital importance to understand their relationship with the pathogenesis of several chronic diseases such as cancer, osteoporosis, Alzheimer's, type 2 diabetes, neurodegenerative diseases, and cardiovascular diseases such as heart failure [7].

The spermatozoon was the first cell type in which the presence of ROS could be identified, because until a few years ago, ROS were considered toxic elements for sperm; however, the RL are currently known (mainly O_2.-) in low concentrations in semen, which play a fundamental role in their biological functions during sperm capacitation, sperm maturation, tyrosine phosphorylation, intergame interaction, and the acrosomal reaction that occurs for fertilization of the oocyte; these phenomena are controlled by the mechanism of defense of enzymatic and nonenzymatic antioxidants that when this balance is broken between the RL and the antioxidant defense system, damages are induced in the nucleic acids, proteins, and lipids present in the membrane of the sperm, causing loss of mobility, decrease in viability, and alterations in the intermediate piece, which finally produce a decrease in seminal quality or sperm death [2, 7, 14, 16, 19–26]. A clear example of this is nitric oxide (NO), which has an important function in the sperm pathophysiology, since in low concentrations it favors the processes of sperm capacitation, the acrosomal reaction, and the union to the zona pelucida; however, in high concentrations it leads to the formation of peroxynitrites, which alters sperm motility [27].

It has been observed that in the ejaculate, the main sources of ROS are leukocytes and abnormal sperm cells, although it has been proposed that there are other possibilities on the generation of intracellular ROS in the spermatozoon, such as the leakage of electrons from the mitochondrial transport chain, NADPH oxidase as a possible source of ROS, and the generation of RL by means of nitric oxide in the post-acrosomal and equatorial regions, which can generate a change in the basal state of the oxidizing agents and induce changes in sperm activity [7].

4. Effect of oxidative stress on the cytoplasmic sperm membrane

The spermatic membrane is asymmetric in its structure and functions. It is formed by an association of phospholipids, plasmalogens, and sphingomyelins in dynamic equilibrium with membrane proteins making it an easy target of oxidizing agents. Cholesterol and phospholipids are important in maintaining the structural integrity of membrane systems. In particular, the plasma membrane of the sperm possesses a large quantity of polyunsaturated fatty acids (PUFA), which are necessary for the acrosome reaction and the interaction with the oocyte membrane. On the other hand, the high content of polyunsaturated fatty acids in the plasma membranes of sperm makes them very susceptible to lipoperoxidation (LP), making it highly vulnerable to oxidative stress [7, 14, 20, 24].

The low concentrations of antioxidant enzymes (catalases, dismutases, peroxidases, and glutathione reductase) in the plasma membrane also convert sperm into cells susceptible to the attack of the RL (particularly the attack of hydroxyl radical (OH) and hydroperoxyl (HO_2)), on all the post-acrosomal region, causing alterations in its permeability (since ROS induces LP of the phospholipids of the membrane, which causes the appearance of "orifices"), affecting the Na^+ and Ca^{2+} pumps, causing these to enter cations into the sperm, altering the osmolarity, which causes the formation of few soluble calcium phosphates, depletion of ATP, and activation by means of Ca^{2+} of proteolytic and phosphoglycolytic enzymes. It also damages the enzymes lactate dehydrogenase, pyruvate kinase, glyceraldehyde 3 phosphate dehydrogenase, and ATPase, generating loss or reduction in mobility, protein and lipid damage, alterations in deoxyribonucleic acid (DNA), anomalies in its morphology, fertility problems, and cell death [9, 14, 20, 23, 24, 28, 29].

5. Effect of oxidative stress on sperm mitochondria

Mitochondria are considered one of the main cellular sources of ROS, which are responsible for regulating physiological processes such as transduction of intracellular signals, the response to oxidative stress, embryonic development, cell proliferation and adhesion, gene expression, and apoptosis [7].

In the sperm mitochondria provide the highest amount of ATP, through glycolysis and oxidative phosphorylation, contributing to the formation of RL during these processes [7, 30, 31]. However, when there is disruption of the mitochondrial respiratory chain (during freezing), these are responsible for the formation and release of ROS. This interruption causes oxygen to undergo complete reductions producing, instead of water molecules, intermediate molecules such as superoxide anion, hydroxyl radical, and hydrogen peroxide, triggering a phenomenon similar to apoptosis, responsible for both the death of sperm and the sublethal damages that decrease the half-life and fertilizing capacity of the cells (**Figure 1**) [32].

The freezing of semen also exerts an important damage in the mitochondria, since it has been demonstrated that the EO induces damage in the mitochondrial DNA, observing that the mutation spectrum of said DNA, in the spermatozoon, can

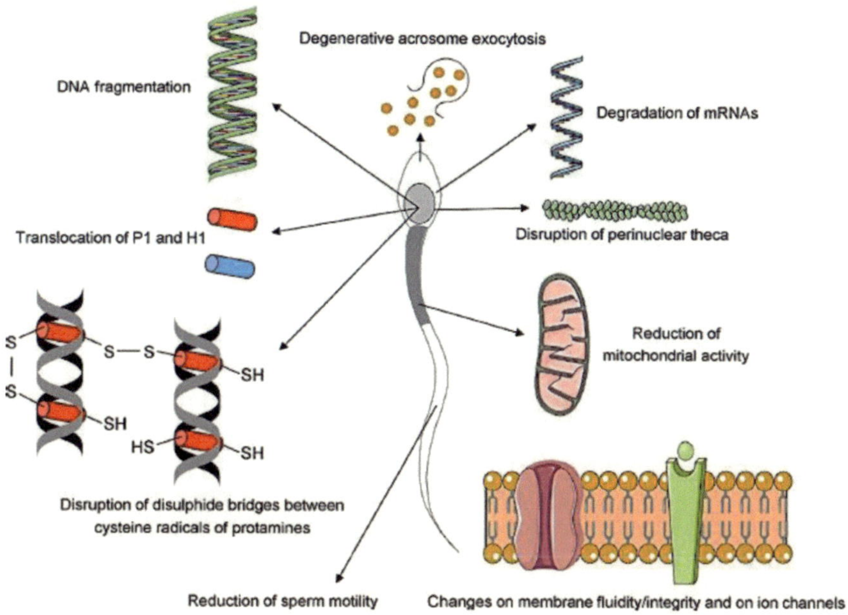

Figure 1.
Lesions resulting from the freezing of pig semen (modified from [4]).

be 10–100 times greater than to nuclear DNA. This can be explained by the cross-linking of DNA proteins that cause RL, exchange of sister chromatids, damage to the structure of deoxyribose phosphate, oxidation of nitrogenous bases, conversion of bases (the deamination of cytosine into uracil and of the 5-methylcytosine in thymidine), ring openings, base release, and chain breaking (one or two strands). This leads directly to a decrease in fertility [4, 7, 9, 24, 33].

6. Effect of oxidative stress on the spermatic acrosome

The acrosome is also affected by the action of the RL during the transport of the sperm through the epididymis, mainly by hydrogen peroxide, since it inhibits the induction of the acrosomal reaction and damages the integrity of the acrosome, producing a malfunction at the time of fertilization of the oocyte [34].

7. Effect of oxidative stress on sperm DNA

Much of the DNA damage in the sperm is generated by the EO. The damage that ROS exerts directly on sperm DNA can induce mutations, affecting the paternal genomics of the embryo, and can be an indication of male fertility [20, 24]. To demonstrate this, in studies where sperm were exposed to high concentrations of artificially produced ROS, a significant increase in DNA damage, decreased sperm motility, and induction in apoptotic processes could be observed [7]. These damages in the chromatic sperm depend on endogenous factors such as in the testicles or the epididymis (during sperm maturation), and exogenous factors as DNA peroxidative damage, infections, immunological factors, or various chemical agents. These may be related to failures in packaging, nuclear maturity, chromatin fragmentation, aneuploidies, or DNA integrity defects [7, 24].

In any part of the spermatogenesis, a damage to the spermatic DNA can be induced, which despite is being a multifactorial phenomenon and not being completely delimited; some of the factors that can produce irreversible damage is the generation of ROS, which come from the respiratory chain, since these oxidative molecules react with the nitrogenous bases and with deoxyribose, causing DNA fragmentation, problems in the compaction and winding of the DNA inside the chromatin, deletions, mutations, translocations, degradation of purine or pyrimidic bases, rupture of chains, and cross-links between proteins and DNA. The magnitude of damage induced by RL during sperm transit through the epididymis depends on the levels of these produced by immature sperm, the presence of epithelial cells or activated leukocytes in the epididymis, and the levels of antioxidant enzymes present in the epididymis lumen [2, 4, 21, 23, 24, 34–37].

It is important to note that there are mainly two RL that affect the DNA strand. The first is the OH radical, which results in the formation of 8-OH-guanine and 8-OH-2 deoxyguanosine at the first stage, attacking the purines as pyrimidines, causing fragmentation of double-stranded DNA, and the second is the radical O21, which generally produces only guanine adducts, especially 8-hydroxyguanine, which affect sperm motility [4, 7, 9, 24]. If a sperm with fragmentation of double-stranded DNA manages to fertilize an oocyte, it is incompatible and may affect the normal development of pregnancy [24].

8. Effect of oxidative stress on female and male fertility

Infertility is defined as the inability of a couple to conceive after a year of sexual intercourse without contraceptive measures [24]. There are multiple causes of male infertility, which may be congenital or acquired; of all of them, idiopathic infertility is caused by multiple factors such as endocrine alterations, oxidative stress, and genetic or epigenetic alterations [38].

In particular, the role of EO as one of the main causes of male infertility has been well established, since ROS can affect all cellular components, including the AGP of membranes, proteins, and nucleic acids, causing in males oligozoospermia, prostate carcinoma, cryptorchidism, varicocele, low seminal quality, low motility of spermatozoa, decreased sperm concentration, and acceleration in the process of apoptosis of geminous cells [24, 27].

In a study conducted by Pérez [27], it was observed that in asthenozoospermic patients have an overexpression of the enzyme inducible nitric oxide synthase (iNOS), compared with the normospermic, which results in a sperm dysfunction and in the decrease of the fecundate capacity of sperm. It has also been shown that in sperm of individuals whose partners have recurrent early embryonic death, there is a significant increase in aneuploidies, abnormal chromatin condensation, DNA fragmentation, apoptosis, and abnormal sperm morphology [19].

It is important to highlight the importance of antioxidants in semen, since it has been observed that the low levels or deficiency of antioxidants in the seminal plasma leaves the sperm unprotected to the EO [20]. So the use of antioxidants has been proposed as a tool to protect sperm from oxidative damage, and it has even been proven that the addition of antioxidants (vitamin C, E or glutathione), at the time of the seminal conservation, produces better results in the seminal evaluation at the time of insemination [4, 7, 29, 39].

In the case of females, it has been suggested that ROS can participate in the formation of adhesions associated with endometriosis, decreasing its fertility. There are also alterations of folliculogenesis caused by ROS, which can deteriorate the quality of the oocyte and have been proposed as a cause of subfertility associated with endometriosis. The EO has also been associated with numerous pathologies among which we can mention mastitis, edema of the udder, higher incidence of diseases in the peripartum period, deficit in the synthesis of steroid horns in cows, and degenerative nutritional myopathy in sheep. In the case of sows, the EO can cause postweaning inflammatory states, modifying the status of selenium and vitamin E affecting the growth rate of piglets [33, 40–42].

9. Conclusions

The effect of EO on sperm cells significantly affects the fecundating capacity of sperm, causing infertility in males and/or low reproductive parameters in females so that the issue of EO in the fertilizing capacity of spermatozoa mammals is of utmost importance at present.

Declaration of conflict of interest

The authors declare that there are no conflicts of interest.

Thanks

To CONACYT, for the support to the second author as a fellow with registration number 624422.

Implications

In this paper, a review is made of a brief history of how the toxicity of oxygen and free radicals was discovered, the oxidative stress in cells, and the effect of oxidative stress in the cytoplasmic sperm membrane, in the spermatic mitochondria, in the spermatic acrosome, in the sperm DNA, and in the fertility of the female and the male.

Author details

Alejando Córdova Izquierdo[1*], Adrian Emmanuel Iglesias Reyes[2], Alda Roció Ortiz Muñiz[3], María de Lourdes Juárez Mosqueda[4], Jesús Alberto Guevara González[1], Juan Eulogio Guerra Liera[5], Gustavo Ruiz Lang[1], Rubén Huerta Crispín[6], Abel E. Villa Mancera[6], Pedro Sánchez Aparico[7] and Raúl Sánchez Sánchez[8]

1 Department of Agricultural and Animal Production, University Autonomous Unit Xochimilco, Mexico City, Mexico

2 Master's Degree in Agricultural Sciences, University Autonomous Unit Xochimilco, Mexico City, Mexico

3 University Autonomous Metropolitan Unit Iztapalapa, Mexico City, Mexico

4 Department of Morphology, FMVZ-UNAM, Mexico City, Mexico

5 Faculty of Agronomy, Autonomous University of Sinaloa, Mexico

6 Faculty of Veterinary, Benemérita Autonomous University of Puebla, Mexico

7 Faculty of Veterinary, Autonomous Mexico State University, Mexico

8 Department of Reproduction, INIA, Madrid, Spain

*Address all correspondence to: acordova@correo.xoc.uam.mx

References

[1] Restrepo BG, Pizarro LE, Albero RB. Estrés oxidativo en el semen equino criopreservado. Revista Lasallista de Investivación. 2012;**9**(1):128-136

[2] Williams S. Criopreservación de semen porcino: Desafíos y perspectivas. Revista Brasileira de Reprodução Animal. 2013;**37**(2):207-2012

[3] Gallardo BJO, Vargas SCA. Evaluacion de tres diluyentes para criopreservar semen bovino de toros cruce Sahiwal (Bos Taurus) En el trópico Húmedo [Tesis previa a la obtención del título de Ingeniero Agropecuario]. Universidad de las Fuerzas Armadas. Departamento de Ciencias de la Vida y la Agricultura; 2015. pp. 1-89

[4] Yeste M, Rodríguez GJE, Bonet S. Artificial insemination with frozen-thawed boar sperm. Molecular Reproduction and Development. 2017;**2017**:1-12

[5] Villa ANA, Ceballos MA. Radicales libres e infertilidad en el macho. Veterinaria e Zootecnia. 2007;**1**(2):87-97

[6] Sánchez I, Torres V, Moreno O, Rodríguez A. Determinación del estrés oxidativo mediante peroxidación lipídica en cristalinos humanos con cataratas. Revista de Facultad de Medicina, Universidad de Los Andes. 2011;**20**(1):42-45

[7] Mayorga TJM, Camargo M, Cadavid Ángela P, Cardona Maya Walter D. Estrés oxidativo: ¿un estado celular defectuoso para la función espermática? Revista Chilena de Obstetricia y Ginecología. 2015;**80**(6):486-492

[8] Martínez SG. Especies reactivas de oxígeno y balance redox, parte 1: Aspectos básicos y principales especies reactivas de oxígeno. Revista Cubana de Farmacia. 2005;**39**(3):1-11

[9] Quintanar EMA, Calderón, Víctor SJ. La capacidad antioxidante total. Bases y Aplicaciones. REB. 2009;**28**(3):89-101

[10] Berzosa S. Estudio del daño oxidativo, niveles de defensa antioxidantes y efecto ergogénico de la melatonina en pruebas de esfuerzo físico agudo [Tesis doctoral para optar al grado de doctor europeo]. Universidad de Zaragoza. Facultad de Medicina. Departamento de Farmacología y Fisiología; 2011. pp. 11-15

[11] García TBE, Saldaña BA, Saldaña GL. El estrés oxidativo y los antioxidantes en la prevención del cáncer. Revista Habanera de Ciencias Médicas. 2012;**12**(2):187-196

[12] Villa NA, Sánchez LE, Ceballos A. Actividad de glutatión peroxidasa y superóxido dismutasa en plasma seminal y sangre en cerdos reproductores. Veterinaria e Zootecnia. 2009;**3**(1):9-51

[13] Gašparovic AC, Lovakovic T, Zarkovic N. Oxidative stress and antioxidants: Biological response modifiers of oxidative homeostasis in cancer. Periodicum Biologorum. 2010;**112**(4):433-439

[14] Flores C, Márquez Y, Vilanova L, Mendoza C. Dienos conjugados y malondialdehído como indicadores de lipoperoxidación en semen de toros "Carora". Revista Veterinaria. 2011;**22**(2):91-94

[15] Villalba MC. Implicaciones del estrés oxidativo en la infertilidad masculina: Análisis de marcadores bioquímicos en plasma seminal y su asociación con parámetros del seminograma y la capacitación espermática [Mediora presentada por Celia Villalba Martínez para optar al grado de Doctor]. Universitat d' Alacant. Departamento de Biotecnología; 2014. pp. 1-321

[16] Gumbao B, David D. Efecto antioxidante del glutatión aplicado en el medio de descongelación seminal de tres especies con interés. Universidad de Murcia. Departamento de Fisiología; 2015. pp. 1-154

[17] Zepeda AB, Farías JB. Antioxidantes frente a estrés oxidativo inducido por hipoxia hipobária en testículo y epidídimo. Revista de Farmacología de Chile. 2013;**6**(1):31-36

[18] Cota MAI. Actividad de las enzimas antioxidantes: Superóxido dismutasa, catalasa y glutatión peroxidasa, en el espermatozoide y líquido seminal de conejo nueva Zelanda y su relación con el sobrepeso [tesis para obtener el grado de Maestra en Biología de la Reproducción Animal]. Universidad Autónoma Metropolitana Unidad Iztapalapa; 2014. pp. 1-71

[19] Rodríguez E, Gil VAM, Aguirre ADC, Cardona MW, Cadavid AP. Evaluación de parámetros seminales no convencionales en individuos cuyas parejas presentan muerte embrionaria temprana recurrente: En busca de un valor de referencia. Biomédica. 2011;**31**:100-107

[20] Villa NA, Castaño D, Duque PC, Ceballos A. Glutatione peroxidase and superoxide dismutase activities in blood and seminal plasma in colombian stallions. Revista Colombiana de Ciencias Pecuarias. 2012;**25**:64-70

[21] Santiani A. Uso de Antioxidantes para mejorar la calidad de semen criopreservado. Spermova. 2013;**3**(2):154-157

[22] Zhong RZ, Zhou DW. Oxidative stress and role of natural plant derived antioxidants in animal reproduction. Journal of Integrative Agriculture. 2013;**12**(10):1826-1838

[23] Orozco BMG, Navarrete MR, Murray NR, Curiel PE. Efecto de la temperatura en el proceso de criopreservación, sobre la motilidad progresiva del espermatozoide de cerdo. Revista Educatecnociencia. 2014;**4**(59):53-64

[24] Paparella CV, Pavesi AB, Feldman RN, Bouvet BR. Importancia de la evaluación del estrés oxidativo en el semen humano. Archivio di Medicina Interna. 2015;**37**(1):7-14

[25] Álvarez Rodríguez M, Vicente Carrillo A, Rodríguez Martínez H. Exogenous individual lecthin-phospholipids (Phosphatidylchoine and Phosphatidylglycerol) cannot prevent the oxidative stress imposed by criopreservation of boar sperm. Journal of Veterinary Medicine and Surgery. 2017;**1**(2):1-11

[26] Domínguez CO, Toledano OÁ, Ávalos RA. Efecto del suplemento de astaxantina sobre la calidad seminal en *Moenkhausia sanctaefiloenae* (Teleostei: Characidae). Latin American Journal of Aquatic Research. 2015;**43**(1):215-221

[27] Pérez MS. Óxido nítrico sintetasa y nitración en tirosina en la astenozoospermia: Un estudio inmunológico. Revista Saegre. 2012;**19**(3):57.59

[28] Pulgar PEA. Sistemas transportadores de vitamina C en células espermatogénicas y espermatozoides [tesis de grado presentada como parte de los requisitos para optar al grado de Licenciado en bioquímica y al título profesional de bioquímico]. Facultad de Ciencias, Universidad Austral de Chile; 2009. pp. 1-192

[29] Thongrueang N, Chaibangyang N, Chanapiwat P, Kaeoket K. Effects of adding melatonin on the quality of frozen-thawed boar semen. Journal of Applied Animal Science. 2017;**10**(2):47-56

[30] Ramón MMO. Estudio de la actividad aminopeptidásica en espermatozoide astenozoospérmicos. Comparación clínica [tesis doctoral]. Universidad del País Vasco, Facultad de Medicina y Odontología, Departamento de Fisiología; 2014. pp. 1-249

[31] Flores C, Vilanova L. Metabolismo espermático. Gaceta de Ciencias Veterinarias. 2015;**20**(1):23-32

[32] Ortega FC. Factores implicados en la variabilidad individual en la respuesta a la congelación del eyaculado equino: Estructura de subpoblaciones, estrés oxidativo y cambios apoptóticos [tesis doctoral]. Universidad de Extremadura, Facultad de veterinaria, Departamento de Medicina Animal; 2011. pp. 1-176

[33] Gupta S, Goldberg Jeffrey M, Aziz N, Goldberg E, Krajcir N, Agarwal A. Mecanismos patogénicos de la infertilidad asociada con endometriosis. Revista Mexicana de Medicina de la Reproducción. 2010;**3**(2):83-97

[34] Córdova JCA. Control de la peroxidación lipídica del semen refrigerado y criopreservado de verraco mediante antioxidantes (α-Tocoferol/ Glutatión reducido) y su repercusión sobre la calidad espermática [tesis doctoral]. Universidad de León. Facultad de Veterinaria. Departamento de Sanidad Animal; 2010. pp. 1-290

[35] Birben E, Murat SU, Sackesen C, Erzyrum S, Kalayci O. Oxidative stress and antioxidant defense. The World Allergy Organization Journal. 2012;**5**(1):9-19

[36] Mayorga TJM, Peña B, Cadavid ÁP, Walter CM. La importancia clínica del ADN espermática en el análisis seminal cotidiano. Revista Chilena de Obstetricia y Ginecología. 2010;**80**(3):256-268

[37] Leyland F. Assessment of ageing-dependent effects on sperm functions following semen cryopreservation. Veterinary Medicine—Open Journal. 2017;**2**(2):1-2

[38] Palma C, Vinay BJI. Infertilidad masculina. Revista Médica Clínica Las Condes. 2014;**25**(1):122-128

[39] Córdova IA, Iglesias RAE, Espinosa CR, Guerra LJE, Villa MAE, Huerta CR, et al. Effect of addition of antioxidants in the extender to freeze boar in two types of straws on sperm quality. International Journal of Recent Scientific Research. 2017;**8**(6):17466-17468

[40] Castro C, Márquez A. Uso de antioxidantes en animales domésticos. Gaceta de Ciencias Veterinarias. 2006;**12**(1):5-12

[41] Quiles A. Efecto del selenio en la producción porcina. Mundo Ganadero. 2008;**8**:42-44

[42] Reinoso V, Soto C. Importancia de la vitamina E y el selenio en vacas lecheras. In: Artigas, Uruguay. Sitio Argentino de Producción Animal, 1-3. 2009. Obtenido de: http://www.produccion-animal.com.ar/suplementacion_mineral/104-Vit_E_y_Se.pdf

Chapter 7

Typical Catalases: Function and Structure

Yonca Yuzugullu Karakus

Abstract

Catalase (EC 1.11.1.6) is a heme-containing enzyme ubiquitously present in most aerobic organisms. Although the full range of biological functions of catalase still remains unclear, its main function is the decomposition of hydrogen peroxide into water and oxygen. Catalases have been studied for over 100 years, with examples of the enzyme isolated, purified, and characterized from many different organisms. The crystal structures of 16 heme-containing catalases have now been solved, revealing a common, highly conserved core in all enzymes. The active center consists of a heme with a tyrosine ligand on the proximal side and a conserved histidine and an aspartate on the distal side. Although catalases have been studied for many years, additional functions of catalases have recently been recognized. For example, *Scytalidium thermophilum* catalase (CATPO) has been shown to oxidize o-diphenolic and some p-diphenolic compounds in the absence of hydrogen peroxide. This and other studies have led to the proposal that this secondary oxidative activity may be a general characteristic of catalases. The present chapter will focus on the function and structure of monofunctional heme catalases, emphasizing the information obtained in the last few years mainly in relation to the secondary activity of these enzymes.

Keywords: catalase, oxidase, heme, NADPH, channel, secondary activity

1. Introduction

Catalases are one of the most studied groups of enzymes. The term catalase was first identified by Loew as hydrogen peroxide (H_2O_2) degrading enzyme in 1901, and the protein has been the focus of study for biochemists and molecular biologists ever since. The overall reaction for catalase can simply be described as the degradation of two molecules of hydrogen peroxide to water and oxygen (reaction 1). This catalytic reaction occurs in two distinct stages, but what each of the stages includes is mainly based on the kind of catalase [1]. The first stage involves oxidation of the heme using first hydrogen peroxide molecule to form an oxyferryl species in which one oxidation equivalent is taken off from the iron and one from the porphyrin ring to make a porphyrin cation radical (reaction 2). In the second stage, this radical intermediate, known as compound I, is reduced by a second hydrogen peroxide to regenerate the resting state enzyme, water and oxygen (reaction 3) [2, 3]. Catalases can also function as peroxidases, in which suitable organic compound is used as an electron donor. During peroxidase reaction, compound I is converted to compound II (reaction 4), which can be oxidized by another hydrogen peroxide to produce the inactive compound III (reaction 5). For NADPH-binding catalases, it has been suggested that enzyme inhibition through the appearance of compound III can be prevented by the NADPH blocking or releasing compound II generation [4–6].

$$2H_2O_2 \rightarrow 2H_2O + O_2 \quad (1)$$

$$\text{Enz}\left(\text{Por}-\text{Fe}^{\text{III}}\right) + H_2O_2 \rightarrow \text{Cpd I}\left(\text{Por}^{+\bullet}-\text{Fe}^{\text{IV}} = O\right) + H_2O \quad (2)$$

$$\text{Cpd I (Por} +\bullet-\text{Fe IV} = O) + H_2O_2 \rightarrow \text{Enz}\left(\text{Por}-\text{Fe}^{\text{III}}\right) + H_2O + O_2 \quad (3)$$

$$\text{Cpd I (Por} +\bullet-\text{Fe IV} = O) + AH_2 \rightarrow \text{Cpd II}\left(\text{Por}-\text{Fe}^{\text{IV}}-\text{OH}\right) + AH^{\bullet} \quad (4)$$

$$\text{Cpd II}\left(\text{Por}-\text{Fe}^{\text{IV}}-\text{OH}\right) + H_2O_2 \rightarrow \text{Cpd III}\left(\text{Por}-\text{Fe}^{\text{III}}-\text{O2}^{-\bullet}\right) + H_2O \quad (5)$$

Catalases have been classified into three groups: monofunctional heme-containing catalases, heme-containing catalase-peroxidases, and manganese-containing catalases [7]. Among them, monofunctional catalases constitute the largest and most extensively studied group of catalases [1, 2]. They all possess two-step mechanism for dismutation of hydrogen peroxide. Members of this largest class of catalases can be biochemically subdivided based on having large (75–84 kDa) subunits with heme *d* associated or small (55–69 kDa) subunits with heme *b* associated. All small subunit enzymes so far characterized, unlike larger enzymes, have been found with NADP(H) bound [1, 8]. In turn, larger subunit enzymes have been shown to exhibit significantly enhanced stability against high temperatures and proteolysis [1, 9]. The catalase-peroxidases, less widespread class, exhibit significant peroxidatic activity in addition to catalytic activity [2]. They are found in bacteria, archaebacteria, and fungi. Catalase-peroxidases have a molecular mass in the range of 120–340 kDa [10, 11]. Manganese-containing catalases are not as widespread as the heme-containing catalases, and there are only three of them so far characterized, one from lactic acid bacteria (*Lactobacillus plantarum*) and two from thermophilic bacteria (*Thermus thermophilus* and *Thermoleophilum album*) [1, 2]. These enzymes are also called pseudo-catalases as their active site contains a manganese-rich reaction instead of heme group [12, 13]. Crystal structures of two manganese catalases, one from *T. thermophilus* and the other from *L. plantarum*, show the presence of dimanganese group in the catalytic center [1].

Although monofunctional catalases are described as such due to the prolonged-agreed belief that their only role is hydrogen peroxide removal, this rather limited catalytic role has recently been questioned. Vetrano et al. expressed a novel oxidase activity in the absence of hydrogen peroxide [14]. Later, a catalase from *S. thermophilum* was shown to have an unselective phenolic oxidase activity in the absence of hydrogen peroxide [15–17]. It is thought that such bifunctional enzymes might be more common due to the evidence on the presence of oxidase/peroxidase activity in catalase enzymes from different organisms such as *Bacillus pumilus* [18], *Thermobifida fusca* [19], and *Amaranthus cruentus* [20]. Such studies are likely to give evidence that translates from various sources to a great deal of catalases. Bifunctional enzymes can be advantageous in many industrial applications including the removal of toxic chemicals and/or chemoprotective agent activity especially when the oxidase activity is enhanced by directed evolution or engineering.

2. Regulation of catalase gene expression

The study of the bacterial response to oxidative stress has given insights into how catalase synthesis is controlled in different cells. Studies with *E. coli* and *Salmonella typhimurium* have shown that there are two regulatory pathways available in bacterial catalase expression [9, 21].

E. coli produces two catalases or hydroperoxidases, the bifunctional catalase-peroxidase HPI and the monofunctional catalase HPII. These two types of catalases are induced independently; HPI synthesis is promoted by H_2O_2 added to a medium, and HPII synthesis is induced during growth into stationary phase [22]. The katG gene, encoding HPI, has been found to be regulated by the OxyR regulon which responds to oxidative stress [9, 21, 22]. OxyR protein is a member of LysR family of regulatory proteins that respond to oxidant levels in the cell [9]. OxyR protein undergoes a conformational change during its transition from the reduced (transcriptionally inactive) to the oxidized (transcriptional active) form. This protein directly senses the oxidative stress by becoming oxidized, and that oxidation results in conformational change by which it transduces oxidative stress to RNA polymerase [21].

The regulatory mechanism of the katE gene, encoding HPII, is quite different and requires a functional katF gene as a positive effector [22]. HPII levels are expressed at high levels when cells enter stationary phase and are unaffected by hydrogen peroxide and/or anaerobiosis [9, 22]. The most important factor for HPII induction seems to be σ^S, as concluded from studies related with the involvement of additional transcription factors [22, 23].

3. Catalase cofactors

The prosthetic group of horse liver catalase enzyme was first isolated by Stern in 1935 [24]. This non-covalently bound component was identified as protoheme (also called hematin), consisting of an iron atom and a porphyrin ring.

The heme prosthetic group has been found to be buried inside the protein, approximately 20 Å from the surface in almost all hem-containing catalases whose structures have been dissolved [25–28]. Despite the similarities in heme-binding pocket, catalases from different sources contain different prosthetic groups [29]. All small subunit size catalases have been shown to include a non-covalently bound iron protoporphyrin IX (heme *b*) as prosthetic group per subunit [29, 30]. Consecutively, an oxidized form of protoporphyrin IX, heme *d*, has been found in almost all large subunit size catalases [30]. The heme *d* group characterized in the active sites of crystal structures of two large subunit size catalases, *Penicillium vitale* catalase (PVC) and HPII from *E. coli*, has the structure of the cis-hydroxy γ-spirolactone and is rotated 180 degrees about the axis defined by the α-γ-meso carbon atoms, with regard to the orientation found for heme *b* in small subunit size catalases like bovine liver catalase (BVC) [29]. **Figure 1** shows the structural differences between *b*-type and *d*-type heme.

The γ-spirolactone ring and additional hydroxyl group make heme *d* more asymmetric with respect to heme *b*. The conversion of heme *b* to heme *d* has been studied in *E. coli* by many scientists, and it is proposed that the oxidation of heme in HPII may be catalyzed by HPII itself. Loewen and colleagues [32] also reported this conversion in the presence of hydrogen peroxide. However, the modification takes place on the proximal side of ring III opposite to the essential distal histidine [29, 33]. Díaz et al. proposed another possible change of protoheme to heme,

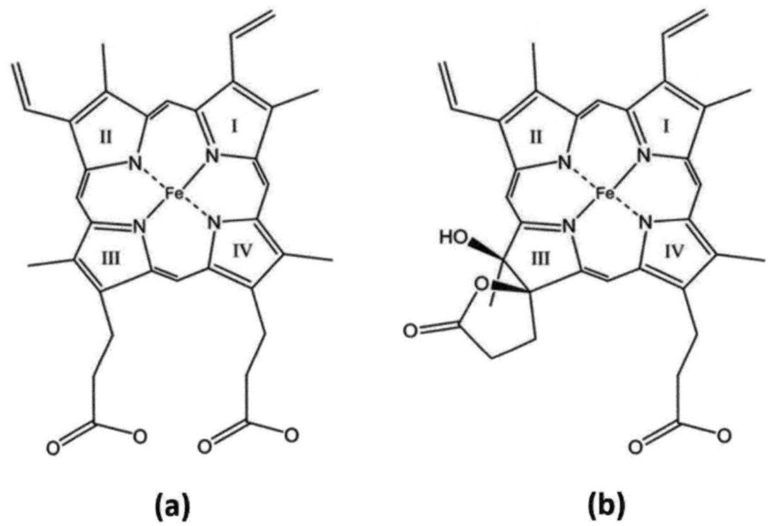

Figure 1.
Structures of heme b (a) and heme d (b), taken from the study reported by Yuzugullu et al. [31].

where γ-spirolactone is formed either by a singlet oxygen or in a light-mediated mechanism [34].

The residues in a contact with heme in the active center are shown to be different for protoheme and heme d enzymes. Such residues for BLC include Met60, Ser216, Leu298, and Met349, whereas analogous residues for PVC involve Ile41, Val209, Pro291, and Leu342 and for HPII contain Ile114, Ile279, Pro356, and Leu407 [29, 35].

Small subunit size catalases have the ability to bind NADP(H) cofactor which is not essential for the activity of catalase [36], but it is believed to have a role in protecting the enzyme from the formation of catalytically inactive intermediate (cpd II) by promoting its reduction to resting state (Fe^{3+}) during catalytic cycle [37, 38]. According to this hypothesis, large subunit enzymes, whose catalytic cycle lacks compound II formation, do not require to bind NADP(H) [38]. It has also been found that NADP(H) is essential for the dismutation of small peroxides, other than hydrogen peroxide [37]. Instead, large subunit size catalases possess the extra C-terminal domain with a flavodoxin-like topology [29, 30]. Despite this difference, residues defining the NADPH pocket in the bovine liver catalase appear to be well preserved in HPII. Only two residues that interact ionically with NADP(H) in the bovine catalase (Asp212 and His304) differ in HPII (Glu270 and Glu362), but it has been proven that their mutation to the bovine sequence does not promote nucleotide binding [4].

4. Catalase catalytic cycle

As described previously, catalytic reaction occurs in two steps [1–3]. The first phase of catalytic cycle involves reaction of ferric enzyme and hydrogen peroxide molecule to generate compound I and water. In the second stage, compound I combines with a second molecule of hydrogen peroxide molecule to regenerate the ferric enzyme, molecular oxygen, and water [2].

Paulos and Kraut firstly proposed the formation of compound I using crystal structure of cytochrome c peroxidase in 1980 [39]. According to this mechanism, proton transfer takes places from hydrogen peroxide to distal imidazole group, and iron-oxygen bond is generated [40]. The studies of water release or rebinding to the coproduct formation site have shown that compound I intermediate might exist in two forms either in a wet form in which a water molecule is present at or near the site of coproduct water formation or dry form where the coproduct water formation site is dry. It is assumed that the presence of water may play a significant role in both substrate selectivity and the variety of redox pathways available in the donor oxidation phase of the catalytic cycles [40, 41].

Compound I intermediate is also perceived in the presence of organic peroxides as substrate, and the reaction rate of compound I production decreases with an increase in the molecular size of the leaving group such as H— > CH_3— > $HOCH_2$— > CH_3C H_2— > $CH_3C(=O)$ — > CH3(CH_2)$_2$— > $CH_3(CH_2)_3OOH$— [42]. At low hydrogen peroxide concentrations and in the presence of suitable organic electron donors, compound I can be reduced by one-electron addition leading to the formation of compound II (a formal Fe^{4+} state) which can cause enzyme inactivation. In this reaction, the porphyrin accepts one electron, therefore losing its radical character [43, 44].

5. Kinetics

The proposed catalytic mechanism supports that catalase enzyme is never saturated with its substrate, H_2O_2, and that turnover of enzyme increases indefinitely as substrate concentration increases [2]. Apparently, catalases have been recognized with a rapid turnover rate and the maximum observed velocities ranging between 54,000 and 833,000 reactions per second [3].

The classical kinetic parameters, Vmax, kcat, and Km, cannot be directly applied to the observed data as catalases do not follow Michaelis-Menten kinetics except at very low substrate concentrations. However, at concentrations below 200 mM, all small subunit size catalases show Michaelis-Menten-like dependence of velocity. At concentrations above 300–500 mM, most small subunit size catalases suffer inactivation. Conversely, large subunit size catalases begin to suffer inhibition above 3 M hydrogen peroxide concentrations [1, 3].

6. Overall structure of catalases

All catalases, whose structure have been dissolved, exhibit highly conserved β-barrel core structure [45]. Their structure is composed of four domains (**Figure 2**) [26, 30, 46, 47]:

a. An amino-terminal arm

b. An anti-parallel eight-stranded β-barrel domain

c. Wrapping domain

d. α-helical domain

The amino-terminal domain is an extended arm and is quite variable in length ranging from 53 residues in *Proteus mirabilis* catalase (PMC) to 127 in HPII [30, 47]. This domain is shown to constitute expanded intersubunit interactions, and residues from this region confer us to describe the heme pocket of a symmetry-associated subunit. The frequency of intersubunit interactions increases with the length of the domain demonstrating catalases' molecular stability [30].

The second domain, referred to as β-barrel domain, is the central feature of catalase. Most of the residues involved in forming the cavity on the distal side of the heme are placed in the first half of the β-barrel. On the other hand, the second half corresponds to the NADP(H)-binding pocket in small subunit catalases. This domain also involves at least six helices situated in two long insertions between β-strands along the polypeptide chain [30, 47].

The wrapping loop is an extended region of almost 110 residues that link the β-barrel and α-helical parts. This region, residues from 366 to 420, does not have any secondary structure except the essential helix (α9) stating the proximal side of heme with tyrosine residue. This part of the polypeptide chain is involved in different interdomain and intersubunit interactions especially with residues from the amino-terminal arm region from another subunit [30, 47].

The α-helical region contains four anti-parallel helices that are close to some of the helices from the β-barrel domain [30, 47].

Unlike BLC, the structures of PVC and HPII present an extra carboxy-terminal domain including roughly 150 residues with a high content of secondary structure elements organized with a "flavodoxin-like" topology [30, 46, 47]. The possible role of this extra domain in PVC remains unknown [30]. In BLC, prior to the flavodoxin-like domain is occupied by an NADP(H) molecule [48].

Although PVC and HPII share common structural similarities, HPII differs in the existence of 60 residues at N-terminal end that increase the contact area between subunits [25].

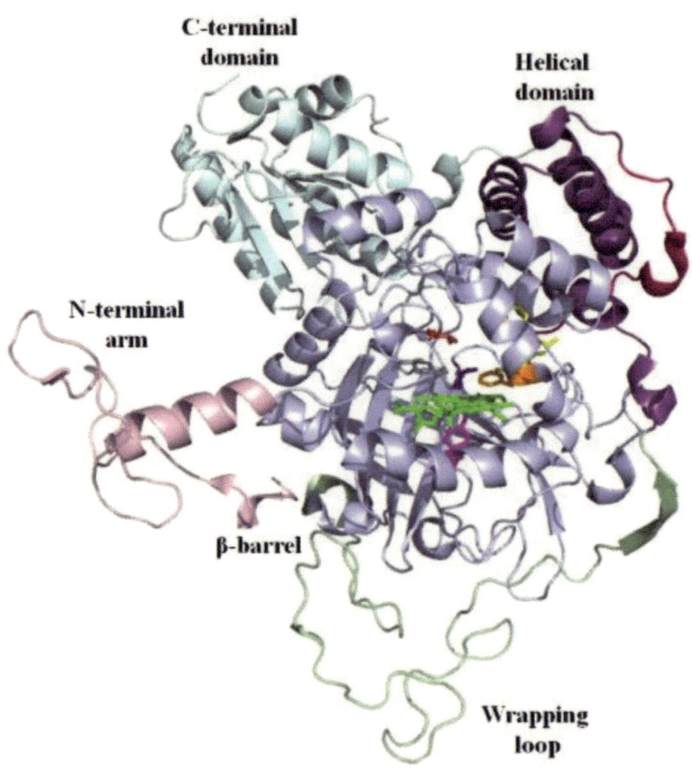

Figure 2.
Schematic drawing of the polypeptide chain and elements of secondary structure in a S. thermophilum *catalase subunit. The heme is colored green, Tyr369 magenta, His82 gray, Asn155 purple blue, Val123 red, Phe160 lemon, Phe161 yellow, and Phe168 orange. This figure is taken from the report of Yuzugullu et al. [17].*

6.1 Heme pocket

In all catalases, the heme group is deeply buried in the core structure, and its distance from the nearest part of the molecular surface is about 20 Å [9, 30]. Three residues, tyrosine on the proximal side of the heme (Tyr415 in HPII) and histidine and asparagine on the distal side (His128 and Asn201 in HPII), are believed to be essential for catalysis [30]. The oxygen of phenolic hydroxyl group in tyrosine residue is the proximal ligand of heme iron and is probably deprotonated with negative charge, so that it can lead to the stabilization of iron's high oxidation states. The imidazole ring of distal histidine is placed almost parallel to the heme at a mean distance of about 3.5 Å above either pyrrole ring III in PMC or pyrrole ring IV in PVC and HPII [9]. The histidine and asparagine residues on the distal side of the heme make the environment strongly hydrophobic [30]. A conserved serine residue (Ser167 in HPII) is also found to be hydrogen bonded to the N^{δ} of the essential histidine and might facilitate the enzymatic mechanism [46].

Despite possessing the same type of heme in active site, PVC and HPII differ in the presence of covalent bond between tyrosine and histidine residues. HPII contains a novel type of covalent bond joining the C^{β} of the essential Tyr415 and the N^{δ} of His392 but not in PVC [33, 44, 46, 49].

6.2 Channels to the heme group

The limited accessibility to heme grouping catalases requires the presence of channels [30]. The heme of the enzyme is connected to the exterior surface by three channels, namely, the main channel, the lateral channel, and the central channel. Among them, the main channel is placed perpendicular to the surface of the heme. The lateral channel approaches horizontal to the heme and the central one heading from the distal side [34, 45].

The main channel is considered to be the primary route for substrate movement to the active site [1, 3]. It is funnel-shaped with 30 Å long in small catalases [30, 48], while in large catalases that channel is replaced by an elongated, constricted, and possibly bifurcated channel that includes the C-terminal domain of adjacent subunit [3, 30].

The conserved residues in the main channel are shown in **Figure 3** including the essential histidine, a valine, and an aspartate (His82, Val123, and Asp135 in CATPO) situated 4, 8, and 12 Å from the heme, respectively [17]. The histidine residue is essential for catalysis in HPII, and the side chain of valine residue makes the channel narrower to a diameter of about 3 Å that prevents any molecule larger than H_2O and H_2O_2 from gaining access to the active site. The role of aspartate has not been investigated in any catalase, but the presence of negatively charged side chain has been found to be critical for catalysis [45].

The lateral or minor channel approaches heme above and below the essential asparagine and emerges in the molecular surface at location corresponding to the NADP(H)-binding pocket in catalases that bind a cofactor (**Figure 4**) [30, 50]. The function of this channel remains unknown [34]. Molecular dynamics analysis indicates that water can exit the protein through this channel [4].

The main channel is a preferred route for substrate entry, but it might be too long and narrow for the release of reaction products (water and molecular oxygen). As the central channel is mainly hydrophilic and leads to the central cavity that is contiguous to the bulk water, this could be a way out for O_2. However, substitutions of amino acid residues extending the major channel in large catalases might allow the exit of oxygen through the main channel. In fact, oxygen preferentially exits

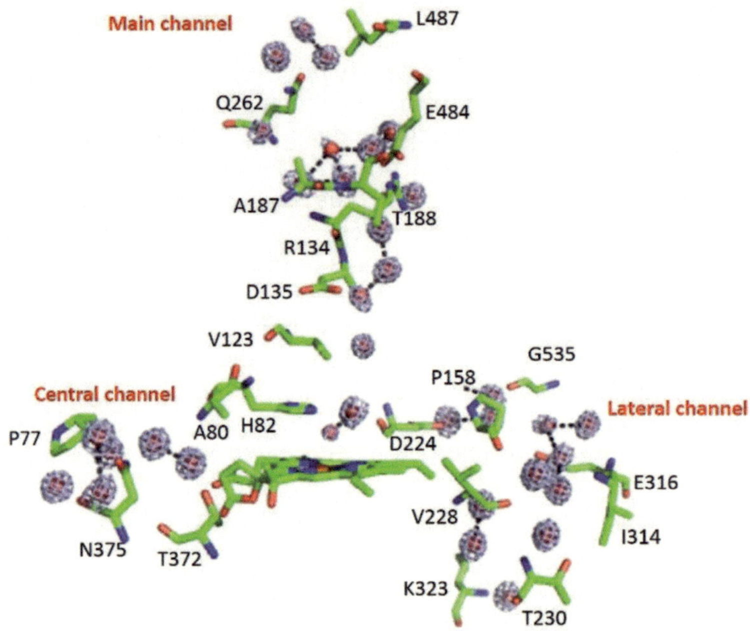

Figure 3.
Channels in CATPO of S. thermophilum.

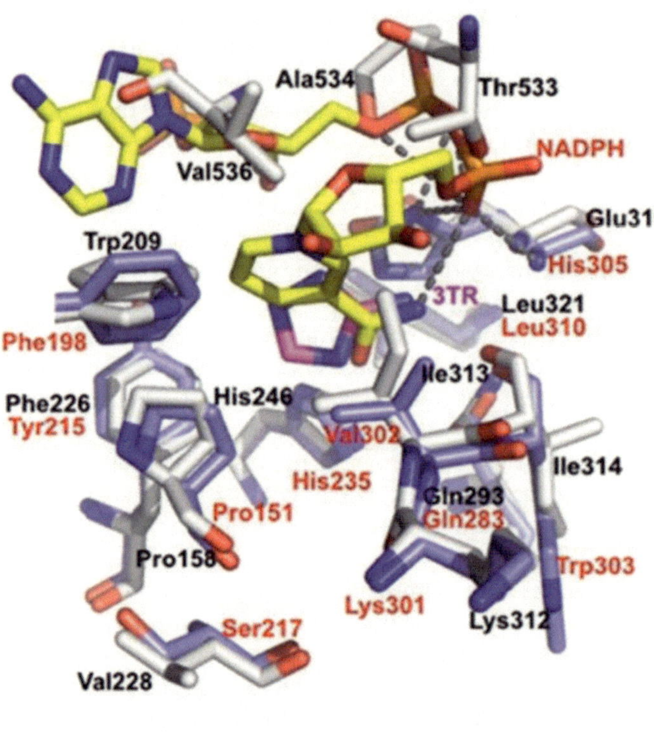

Figure 4.
View of chain A of CATPO complex with 3TR (PDB 5ZZ1, gray) superposed onto human catalase (PDB 1DGH7, blue). CATPO loop 533–537 lies across the top of the NADPH-binding pocket, clashing with the position of the NADPH in the human enzyme [50].

through the main channel instead of central one in all catalases having b-type heme in the active site. Thus, the presence of minor channels might be an alternative mechanism for a fast release of products under the condition of high H_2O_2 stress. These results indicate that O_2 can exit the enzyme through different channels although the main exit in large catalases might be through the central channel and in small catalases through the major channel [34, 51].

6.3 Bifunctionality of catalase and phenol oxidase

Many reports on catalase and phenol oxidase enzymes suggest that the activities may flap in some way that catalases exhibit additional oxidase activity and phenol oxidases present further catalase activity. This relationship can be explained by the release of H_2O_2 due to polyphenol oxidation [52]. Hydrogen peroxide generation by phenol oxidation was also reported by Aoshima and Ayebe [53]. They observed high concentrations of H_2O_2 in beverages like tea or coffee directly after opening caps as a result of oxygen.

Jolley et al. first developed mushroom tyrosinase with catalase activity in the presence of hydrogen peroxide [54]. Garcia-Molina et al. [55] and Yamazaki et al. [56] also studied this bifunctional behavior of tyrosinase. In addition to this novel tyrosinase, a catalase-like process was found to have one isozyme of catechol oxidase from sweet potatoes (*Ipomoea batatas*) [57].

In literature, the first report on catalase known as a monofunctional enzyme but possessing secondary activity (oxidase) was introduced for mammalian catalase. This enzyme has been reported to present oxidase activity when hydrogen peroxide is absent or levels of H_2O_2 are low. As mentioned previously, the main function of catalase is the decomposition of hydrogen peroxide into water and oxygen (catalytic activity). Moreover, it is known that catalases can oxidize low molecular weight alcohols in the presence of low concentrations of H_2O_2 (peroxidatic activity). The catalytic mechanism of catalases is a two-step process in which catalase heme Fe^{3+} reduces one hydrogen peroxide molecule to water and generates a porphyrin cation radical called compound I, which is then oxidized by a second hydrogen peroxide to give molecular oxygen and water. The peroxidase activity stems from the oxidation of alcohols by compound I through single-electron transfer. Vetrano et al. expressed a novel oxidase activity in the absence of hydrogen peroxide. This oxidase reaction involves the interaction of catalase heme with a strong reducing agent like benzidine (HB) and molecular oxygen leading to the formation of a compound II-like intermediate. The subsequent electron transfer causes substrate oxidation and regeneration of resting enzyme. An incomplete reaction may result in the formation of radical centered intermediates and the production of superoxide [14].

Later, catalase from the thermophilic fungus *S. thermophilum* has been reported to possess additional phenol oxidase activity [16]. This enzyme, named as CATPO, is the first bifunctional catalase-phenol oxidase in the literature that is characterized in detail. *S. thermophilum* CATPO is a homotetramer with a molecular mass of 320 kDa. Based on the amino acid sequence and preliminary three-dimensional structure [58], CATPO is classified as a large heme catalase with the highest structural homology (77%) to catalase of *Penicillium vitale* [16]. CATPO can oxidize *o*-diphenols such as catechol, caffeic acid, and L-DOPA in the absence of hydrogen peroxide, and the highest oxidase activity is observed against catechol. This enzymatic activity is oxygen-dependent and is inhibited by classic catalase inhibitors, including 3-amino-1,2,4-triazole (3TR). The peroxide-independent secondary activity has also been identified in other catalases [14, 19, 20] and has been presumed to also occur at the heme active site.

There are a great number of reports available describing the structural and biochemical characterization of catalases. However, basic questions related to substrate and product flow remain unanswered, particularly related to the oxidase activity. Therefore researchers have recently focused on the investigation of the region of CATPO that corresponds to the NADPH-binding region of bovine liver catalase (BLC) and the lateral channel. A number of mutations were introduced into this region, and the properties of these mutant variants, including their specific activities and sensitivities to various inhibitors, are interpreted in terms of a role for the lateral channel in CATPO. The structural, mutation, and kinetic evidences suggested that this pocket at the entrance to the lateral channel, captured by NADPH's nicotinamide moiety in mammalian catalases, should be the site of both oxidase substrate and 3TR binding. The promiscuous nature of CATPO oxidase is clarified by the presence of numerous ordered water molecules which facilitate substrate binding through hydrogen bond formation and can be transferred to accommodate various size and shaped substrates. Peroxide-independent phenolic substrate oxidation is then likely to happen in a similar manner to NADPH oxidation, by electron transfer from the substrate to a high-valent iron-oxo intermediate, apparently arisen through reaction with oxygen [50].

7. Conclusions

Catalases have been studied for over 100 years, with examples of the enzyme isolated, purified, and characterized from many different organisms. The crystal structures of 16 monofunctional catalases have been solved at high resolution. These structures show that they are tetramers, and each of the four active sites consists of a pentacoordinated-iron protoporphyrin IX prosthetic group with a tyrosinate axial ligand. Some also contain a NADPH cofactor tightly bound at the periphery of each subunit. Recently, it has been found that these enzymes exhibit an oxidase activity in addition to their H_2O_2 degrading activity. Although they are old enzymes, a peroxide-independent oxidase activity of catalases is new in the literature. Such studies have led to the proposal that this secondary oxidative activity may be a general characteristic of catalases.

Acknowledgements

The author is grateful to the Department of Biotechnology, Middle East Technical University, Turkey, for providing the necessary facilities during PhD program and many thanks to the Department of Biology, Kocaeli University and TUBİTAK (grant No. 113Z744) for financial support. The author is also grateful to the EU COST Action CM1306 "Movement and Mechanism in Molecular Machines" for STSM support.

Conflict of interest

The authors declare no conflict of interest.

Appendices and nomenclature

BLC	bovine liver catalase
CATPO	catalase-phenol oxidase
Cpd	compound

3TR	3-amino-1,2,4-triazole
H_2O_2	hydrogen peroxide
HB	benzidine
HPI	hydroperoxidase I
HPII	hydroperoxidase II
L-DOPA	L-3,4-dihydroxy-phenylalanine
NADP(H)	nicotinamide adenine dinucleotide phosphate
PVC	*Penicillium vitale* catalase
PMC	*Proteus mirabilis* catalase
POR	porphyrin

Author details

Yonca Yuzugullu Karakus
Department of Biology, Faculty of Arts and Sciences, Kocaeli University, Kocaeli, Turkey

*Address all correspondence to: yonca.yuzugullu@kocaeli.edu.tr

References

[1] Chelikani P, Fita I, Loewen PC. Diversity of structures and properties among catalases. CMLS Cellular and Molecular Life Sciences. 2004;**61**:192-208. DOI: 10.1007/s00018-003-3206-5

[2] Nicholls P, Fita I, Loewen PC. Enzymology and structure of catalases. Advances in Inorganic Chemistry. 2001;**51**:51-106. DOI: 10.1016/S0898-8838(00)51001-0

[3] Switala J, Loewen PC. Diversity of properties among catalases. ABB Achieves of Biochemistry and Biophysics. 2002;**401**:145-154. DOI: 10.1016/S0003-9861(02)00049-8

[4] Sevinc MS, Mate MJ, Switala J, Fita I, Loewen PC. Role of the lateral channel in catalase HPII of *Escherichia coli*. Protein Science. 1999;**8**:490-498. DOI: 10.1110/ps.8.3.490.

[5] Putnam CD, Arvai AS, Bourne Y, Tainer JA. Active and inhibited human catalase structures: Ligand and NADPH binding and catalytic mechanism. Journal of Molecular Biology. 2000;**296**:295-309. DOI: 10.1006/jmbi.1999.3458

[6] Nicholls P. Classical catalase: Ancient and modern. Achieves of Biochemistry and Biophysics. 2012;**525**:95-101. DOI: 10.1016/j.abb.2012.01.015

[7] Loncar N, Fraaije MW. Catalases as biocatalysts in technical applications: Current state and perspectives. Applied Microbiology and Biotechnology. 2015;**99**:3351-3357. DOI: 10.1016/j.abb.2012.01.015

[8] Klotz MG, Loewen PC. The molecular evolution of catalytic hydroperoxidases: Evidence for multiple lateral transfer of genes between Prokaryota and from Bacteria into Eukaryota. Molecular Biology and Evolution. 2003;**20**:1098-1112. DOI: 10.1093/molbev/msg129.

[9] Loewen PC. Bacterial catalases. In: Scandalios JG, editor. Oxidative Stress and the Molecular Biology of Antioxidant Defences. Cold Spring Harbor: New York; 1997. pp. 273-308

[10] Nagy JM, Cass AE, Brown KA. Purification and characterization of recombinant catalase–peroxidase, which confers ionized sensitivity in *Mycobacterium tuberculosis*. Journal of Biological Chemistry. 1997;**272**:31265-31271. DOI: 10.1074/jbc.272.50.31265

[11] Obinger C, Regelsberger G, Strasser G, Burner U, Peschek GA. Purification and characterization of a homodimeric catalase–peroxidase from the cyanobacterium Anacystis nidulans. Biochemical and Biophysical Research Communications. 1997;**235**:545-552. DOI: 10.1006/bbrc.1997.6847

[12] Allgood GS, Perry JJ. Characterization of a manganese-containing catalase from the obligate thermophile *Thermoleophilum album*. Journal of Bacteriology. 1986;**168**:563-567. DOI: 10.1128/jb.168.2.563-567.1986

[13] Whittaker MM, Barynin VV, Antonyuk SV, Whittaker JW. The oxidized (3,3) state of manganese catalase: Comparison of enzymes from *Thermus thermophilus* and *Lactobacillus pantarum*. Biochemistry. 1999;**38**:9126-9136. DOI: 10.1021/bi990499d

[14] Vetrano AM, Heck DE, Mariano TM, Mishin V, Laskin DL, Laskin JD. Characterization of the oxidase activity in mammalian catalase. The Journal of Biological Chemistry. 2005;**280**:35372-35381. DOI: 10.1074/jbc.M503991200

[15] Ögel ZB, Yüzügüllü Y, Mete S, Bakir U, Kaptan Y, Sutay D, et al.

Production, properties and application to biocatalysis of a novel extracellular alkaline phenol oxidase from the thermophilic fungus *Scytalidium thermophilum*. Applied Microbiology and Biotechnology. 2006;**71**:853-862. DOI: 10.1007/s00253-005-0216-2

[16] Sutay Kocabas D, Bakir U, Phillips SEV, McPherson MJ, Ogel ZB. Purification, characterization, and identification of a novel bifunctional catalase-phenol oxidase from *Scytalidium thermophilum*. Applied Microbiology and Biotechnology. 2008;**79**:407-415. DOI: 10.1007/s00253-008-1437-y

[17] Yuzugullu Y, Trinh CH, Smith MA, Pearson AR, Phillips SEV, Sutay Kocabas D, et al. Structure, recombinant expression and mutagenesis studies of the catalase with oxidase activity from *Scytalidium thermophilum*. Acta Crystallographica Section D. 2013;**69**:398-408. DOI: 10.1107/S0907444912049001

[18] Loewen PC, Villanueva J, Switala J, Donald LJ, Ivancich A. Unprecedented access of phenolic substrates to the heme active site of a catalase: Substrate binding and peroxidase-like reactivity of *Bacillus pumilus* catalase monitored by X-ray crystallography and EPR spectroscopy. Proteins. 2015;**83**:853-866. DOI: 10.1002/prot.24777

[19] Loncar N, Fraaije MW. Not so monofunctional-a case of the thermostable *Thermobifida fusca* catalase with peroxidase activity. Applied Microbiology and Biotechnology. 2015;**99**:2225-2232. DOI: 10.1007/s00253-014-6060-5

[20] Chen N, Teng X-L, Xiao X-G. Subcellular localization of a plant catalase-phenol oxidase, AcCATPO, from Amaranthus and identification of a non-canonical peroxisome targeting signal. Frontiers in Plant Science. 2017;**8**:1-11. DOI: 10.3389/fpls.2017.01345

[21] Storz G, Tartaglia LA. OxyR: A regulator of antioxidant genes. The Journal of Nutrition. 1992;**122**:627-630. DOI: 10.1093/jn/122.suppl_3.627

[22] Mulvey MR, Switala J, Borys A, Loewen PC. Regulation of transcription of katE and katF in *Escherichia coli*. Journal of Bacteriology. 1990;**172**:6713-6720. DOI: 10.1128/jb.172.12.6713-6720.1990

[23] Meir E, Yagil E. Regulation of *Escherichia coli* catalases by anaerobiosis and catabolite repression. Current Microbiology. 1990;**20**:139-144. DOI: 10.1007%2FBF02091988

[24] Stern KG. The constitution of the prosthetic group of catalase. Journal of Biological Chemistry. 1936;**114**:473

[25] Bravo J, Verdaguer N, Tormo J, Betzel C, Switala J, Loewen PC, et al. Crystal structure of catalase HPII from *Escherichia coli*. Structure. 1995;**3**:491-502. DOI: 10.1016/s0969-2126(01)00182-4

[26] Fita I, Silva AM, Murthy MRN, Rossmann MG. The refined structure of beef liver catalase at 2.5 Å resolution. Acta Crystallogrophica—Section B: Structural Crystallography and Crystal Chemistry. 1986;**42**:497-515. DOI: 10.1107/S0108768186097835

[27] Gouet P, Jouve HM, Dideberg O. Crystal structure of *Proteus mirabilis* PR catalase with and without bound NADPH. Journal of Molecular Biology. 1995;**249**:933-954. DOI: 10.1006/jmbi.1995.0350

[28] Murshudov GN, Melik-Adamyan WR, Grebenko AI, Barynin VV, Vagin AA, Vainshtein BK, et al. Three-dimensional structure of catalase from *Micrococcus lysodeikticus* at 1.5 Å resolution. FEBS Letters. 1992;**312**:127-131. DOI: 10.1016/0014-5793(92)80919-8

[29] Murshudov GN, Grebenko AI, Barynin V, Dauter Z, Wilson KS, Vainshtein BK, et al. Structure of the heme d of *Penicillium vitale* and *Escherichia coli* catalases. The Journal of Biological Chemistry. 1996;**271**:8863-8868. DOI: 10.1074/jbc.271.15.8863.

[30] Maté MJ, Murshudov G, Bravo J, Melik-Adamyan W, Loewen PC, Fita I. Heme catalases. In: Messerschmidt A, Huber R, Poulos T, Widghardt K, editors. Handbook of Metalloproteins. Chichester, UK: Wiley & Sons; 2001. pp. 486-502

[31] Yuzugullu Y, Zengin M, Balci S, Goc G, Avci Duman Y. Role of active site residues on catalytic activity of catalase with oxidase activity from *Scytalidium thermophilum*. Procedia—Social and Behavioral Sciences. 2015;**195**:1728-1735. DOI: 10.1016/j.sbspro.2015.06.289

[32] Loewen PC, Switala J, von Ossowski I, Hillar A, Christie A, Tattrie B, et al. Catalase HPII of *Escherichia coli* catalyzes the conversion of protoheme to cis-heme d. Biochemistry. 1993;**32**:10159-10164. DOI: 10.1021/bi00089a035

[33] Bravo J, Fita I, Ferrer JC, Ens W, Hillar A, Switala J, et al. Identification of a novel bond between a histidine and the essential tyrosine in catalase HPII of *Escherichia coli*. Protein Science. 1997;**6**:1016-1023. DOI: 10.1002/pro.5560060507

[34] Díaz A, Muñoz-Clares RA, Rangel P, Valdés VJ, Hansberg W. Functional and structural analysis of catalase oxidized by singlet oxygen. Biochimie. 2005;**87**:205-214. DOI: 10.1016/j.biochi.2004.10.014

[35] Maj M, Loewen PC, Nicholls P. *E. coli* HPII catalase interaction with high spin ligands: Formate and fluoride as active site probes. Biochimica et Biophysica Acta. 1998;**1384**:209-222. DOI: 10.1016/s0167-4838(97)00167-2

[36] Kirkman HN, Gaetani GF. Catalase: A tetrameric enzyme with four tightly bound molecules of NADPH. Proceedings of the National Academy of Sciences. 1984;**81**:4343-4347. DOI: 10.1073/pnas.81.14.4343

[37] Cattani L, Ferri A. The function of NADPH bound to catalase. Journal of Biological Research-Italian Society for Experimental Biology. 1994;**70**:75-82

[38] Hillar A, Nicholls P. A mechanism for NADPH inhibition of catalase compound II formation. FEBS Letters. 1992;**314**:179-182. DOI: 10.1016/0014-5793(92)80969-n

[39] Poulos TL, Kraut J. The stereochemistry of peroxidase catalysis. Journal of Biological Chemistry. 1980;**255**:8199-8205

[40] Jones P, Dunford HB. The mechanism of compound I formation revisited. Journal of Inorganic Biochemistry. 2005;**99**:2292-2298. DOI: 10.1016/j.jinorgbio.2005.08.009

[41] Jones P. Roles of water in heme peroxidase and catalase mechanisms. The Journal of Biological Chemistry. 2001;**276**:13791-13796. DOI: 10.1074/jbc.M011413200

[42] Hara I, Ichise N, Kojima K, Kondo H, Ohgiya S, Matsuyama H, et al. Relationship between the size of the bottleneck 15 Å from iron in the main channel and the reactivity of catalase corresponding to the molecular size of substrates. Biochemistry. 2007;**46**:11-22. DOI: 10.1021/bi061519w

[43] Alfonso-Prieto M, Borovik A, Carpena X, Murshudov G, Melik-Adamyan W, Fita I, et al. The structures and electronic configuration of compound I intermediates of *Helicobacter*

pylori and *Penicillium vitale* catalases determined by X-ray crystallography and QM/MM density functional theory calculations. Journal of American Chemical Society. 2007;**129**:4193-4205. DOI: 10.1021/ja063660y

[44] Maté MJ, Zamocky M, Nykyri LM, Herzog C, Alzari PM, Betzel C, et al. Structure of catalase A from *Saccharomyces cerevisiae*. Journal of Molecular Biology. 1999;**286**:135-139. DOI: 10.1006/jmbi.1998.2453.9

[45] Chelikani P, Carpena X, Fita I, Loewen PC. An electrical potential in the access channel of catalases enhances catalysis. The Journal of Biological Chemistry. 2003;**278**:31290-31296. DOI: 10.1074/jbc.M304076200

[46] Bravo J, Mate MJ, Schneider T, Switala J, Wilson K, Loewen PC, et al. Structure of catalase HPII from *Escherichia coli* at 1.9 Å resolution. Proteins. 1999;**34**:155-166. DOI: 10.1002/(SICI)1097-0134(19990201)34:2<155:AID-PROT1>3.0.CO;2-P

[47] Melik-Adamyan WR, Barynin VV, Vagin AA, Borisov VV, Vainshtein BK, Fita I, et al. Comparison of beef liver and *Penicillium vitale* catalases. Journal of Molecular Biology. 1986;**188**:63-72. DOI: 10.1016/0022-2836(86)90480-8

[48] Fita I, Rossmann MG. The active center of catalase. Journal of Molecular Biology. 1985;**185**:21-37. DOI: 10.1016/0022-2836(85)90180-9

[49] Melik-Adamyan W, Bravo J, Carpena X, Switala J, Maté MJ, Fita I, et al. Substrate flowing catalases deduced from the crystal structures of active site variants of HPII from *Escherichia coli*. Proteins: Structure, Function, and Genetics. 2001;**44**:270-281. DOI: 10.1002/prot.1092

[50] Yuzugullu Karakus Y, Goc G, Balci S, Yorke BA, Trinh CH, McPherson MJ, et al. Identification of the site of oxidase substrate binding in the *Scytalidium thermophilum* catalase. Acta Crystallographica Section D. 2018;**74**:979-985. DOI: 10.1107/S2059798318010628

[51] Kalko SG, Gelpí JL, Fita I, Orozco M. Theoretical study of the mechanisms of substrate recognition by catalase. Journal of American Chemical Society. 2001;**123**:9665-9672. DOI: 10.1021/ja010512t

[52] Akagawa M, Shigemitsu T, Suyama K. Production of hydrogen peroxide by polyphenols and polyphenol-rich beverages under quasi-physiological conditions. Bioscience, Biotechnology and Biochemistry. 2003;**67**:2632-2640. DOI: 10.1271/bbb.67.2632

[53] Aoshima H, Ayabe S. Prevention of the deterioration of polyphenol-rich beverages. Food Chemistry. 2007;**100**:350-355. DOI: 10.1016/j.foodchem.2005.09.052

[54] Jolley RL Jr, Evans LH, Makino N, Mason HS. Oxytyrosinase. The Journal of Biological Chemistry. 1974;**249**:335-345

[55] Garcia-Molina F, Hiner ANP, Fenoll LG, Rodriguez-Lopez JN, Garcia-Ruiz PA, Garcia-Canovas F, et al. Mushroom tyrosinase: Catalase activity, inhibition and suicide inactivation. Journal of Agricultural and Food Chemistry. 2005;**53**:3702-3709. DOI: 10.1021/jf048340h

[56] Yamazaki S, Morioka C, Itoh S. Kinetic evaluation of catalase and peroxygenase activities of tyrosinase. Biochemistry. 2004;**43**:11546-11553. DOI: 10.1021/bi048908f

[57] Gerdemann C, Eicken C, Magrini A, Meyer HE, Rompel A, Spener F, et al. Isoenzymes of Ipomoea batatas catechol oxidase differ in catalase-like